# San Juan Gold:

## A Mining Engineer's Adventures, 1879-1881

### Duane A. Smith

WESTERN REFLECTIONS PUBLISHING COMPANY®

Montrose, CO

First Edition
Printed in the United States of America

Library of Congress Catalog Card Number 2002100711

ISBN 1-890437-67-0

Cover and text design by Laurie Goralka Design

Western Reflections Publishing Company®
P.O. Box 1647
Montrose, CO 81402
www.westernreflectionspub.com

*For John K. Creighton*
*Friend, fellow historian and running buddy*

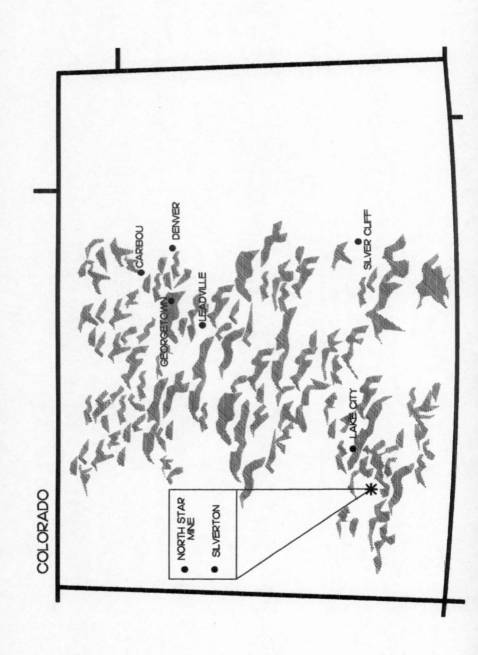

COLORADO

CARIBOU
DENVER
GEORGETOWN
LEADVILLE
SILVER CLIFF
LAKE CITY

NORTH STAR MINE
SILVERTON

# TABLE OF CONTENTS

# PREFACE

*H*iking the old mule trail to the North Star Mine is spectacular in itself. Either the sweeping views or the altitude, or both, take one's breath away. In July 2001, nature offered an extraordinary treat—a wondrous flower display, featuring hillsides decked with columbines, edged by Indian paint brush, against a deep blue Colorado sky. When one finally tops the ridge that is King Solomon Mountain, a San Juan panorama unfolds in all directions.

It is no wonder that Eben Olcott spent so much time trying to picture for his New York family what he was seeing and experiencing. The grandeur of nature was, and is, majestic. Words alone cannot do them justice. The majesty of the scene simply overwhelms the viewer.

While the mine and the era that Olcott knew are gone, the mountains remain. A few boards, a boiler, handfuls of square nails, broken glass, and rods used to anchor the buildings onto the mountainside say that man once worked here, but now nature has reclaimed much of the territory. Those rods and, occasionally along the trail, bent mine rails sunk into the mountain with cables dangling down between them mutely testify to the struggle to travel and to work at 12,800 feet. Gold and silver have a way of luring people into extraordinary effort and struggling to work at amazing places.

Danger existed in all of this. Those rails had once been straight. Rocks and snow, over time, beat down man's iron. And what they did not threaten, the wind did. Working and living at this altitude is an experience in itself. Altitude sickness, headaches, tiring quickly, shortness of breath, rapid changes in weather, extreme conditions—the miners and their manager knew them in detail. Olcott was aware of all of these things and wrote about them to his family, although he spared some aspects so as not to worry them.

A debt of gratitude is owed to many in the trail that led to Olcott. Gene Gressley first called attention to the letters at the University of Wyoming's Heritage Center. A long time colleague, Harry Rosenberg, spent time at the New-York Historical Society ferreting out the material there. Ralph Shearer used his skill at copying photographs to reproduce the nineteenth- century ones which do so much to bring back that lost era. The Homer of the San Juans, Allen Nossaman, graciously contributed his knowledge to the project.

Three friends—Glen Crandall, Shane Voss, and John Ninnemann—hiked up that steep trail to King Solomon Mountain to savor what Olcott lovingly did. They not only partook in the experience but brought back down photographs which are included in this volume. My wife Gay read the manuscript and made her usual pertinent editorial additions. Nevertheless, enough of the present. Journey now back to an earlier era, though not long ago in the history of man.

## PROLOGUE

*H*igh up on King Solomon Mountain in the craggy San Juan Mountains of southwestern Colorado, perches the North Star Mine. Perhaps, following the old mining adage that the higher up the silver mine, the richer the silver, its discoverers expected a bonanza. Those that followed them certainly did. The North Star, at more than 12,800 feet, was elevated, even for the San Juans, themselves the highest mining district in the United States.

The San Juans had been tempting prospectors and miners since the days of the Spanish.

The veins that crisscrossed the mountains seemed to imply, to the hopeful, the eager, and the adventuresome, that here existed the real mother lode country where vast riches awaited the fortunate. In the eighteenth century, the Spanish arrived from New Mexico leaving behind names—San Juan, La Plata, Animas—and legends of "lost" mines. The isolation, the Utes, the rugged terrain, and a challenging environment defeated all efforts except dreams.

The 1849 California gold rush turned dreams into reality and, a decade later, the Pike's Peak rush brought the hopeful to the central Rocky Mountains. In the next year, 1860, Charles Baker led a party up the Animas River into what became Baker's Park, where Silverton one day would be located. His enthusiastic promotion produced an ill-fated rush, the next year, into the heart of the San Juans. The major problem confronted by one and all: gold did not reward the rushers' determined efforts and isolation dogged their every step. Defeated, they retreated and not until the early 1870s would they return.

The dreams of wonderful riches hidden in the San Juans did not die. Back came the prospectors, despite all the problems. As it was said of miners, they would cross heaven to get to hell, if there was a rumored gold strike in Hades. This time, after an initial few seasons of late spring to early fall prospecting and digging, permanent settlement took root. From Del Norte, they labored westward over the well-named Stony Pass and down Cunningham Gulch to the Animas River.

Along the San Juans' eastern fringe Lake City became a gateway to the little camps located beyond across the mountains. Over Engineer and Cinnamon passes struggled excited throngs to seek their fortunes. They passed the hamlets of Capitol City, and Rose's Cabin along the former. Sherman, Burrows Park and Tellurium set astride the other, each offering a few amenities to the travelers. Urbanization, so typical of a mining rush, had come to stay in the San Juans.

Nestled by the banks of the Animas River, curving northward from Silverton were Howardsville, Middleton, Eureka, and Animas Forks, tiny villages vital to the existence of the miners working beyond them. By the time prospectors climbed King Solomon Mountain and reached the future North Star Mine, the district had become known, and year-around mining was possible at lower elevations with the help of the support hamlets in the valleys. That did not lessen the region's isolation, transportation problems, and robust climate. As one miner emphatically resolved, "he would be damned if he would stay in a place with three months of mighty late fall and nine months of damned hard winter." That might be an apocryphal utterance, but it was true, nevertheless.

More significant to the future and well being of the district were the observations of United States Commissioner of Mining Statistics, Rossiter Raymond. He wrote, in his 1875 report, that much had been accomplished in the last two years, "notwithstanding the great distance from railroad communications, the inaccessibility of the country, and the want of capital for the development of a new mining-camp."

The soon-to-be centennial state of Colorado had not been popular with the investing public since the fiasco of the busted gold mining stock speculation in 1863-64. Lost investments, financial shenanigans, over-expectations, and failed mining methods rained damnation on the young territory. That hurt the San Juans, even a decade later. A newly opening district needed three things to make it successful. Raymond touched upon two that the San Juans desperately craved, but had not secured— investment, and the best possible transportation, which translated for late nineteenth century Americans to the railroad. The third, the mountains possessed in abundance-rich ore.

The region had gained the urban base that mining needed. Lake City and Silverton emerged as the two most prominent camps. Lake City, with nearby accessible rich surface deposits, an easy-to-reach location, and the gateway to two of the passes into the heart of the mountains, jumped to an early lead. It also suffered an early slump as the local ore did not maintain its promise. It was more tantalizing than anything else. Denver's *Rocky Mountain News* (June 22, 1879) blamed its problems on too little money, weak faith, and apathy.

From its birth in 1873, Silverton struggled with neighboring Howardsville to gain supremacy over the southern part of the San Juans. It won, capturing the county seat, smelting facilities, and potentially rich mines surrounding Baker's Park. The *Ouray Times* (July 27, 1878) described Silverton as "quite lively at present" with considerable building going on and "full" hotels. Census takers in 1880 counted 607 people in "Lake town," and 1,151 in Silverton.

Meanwhile, prospectors had been scurrying around Silverton's Animas district and toiling up "silver-ribbed" King Solomon Mountain. Separating Cunningham Gulch from the Little Giant Basin (where the mine of the same name became the San Juans' first boom and bust), King Solomon's "lofty summit" topped 13, 185 feet. As the early 1870s gold rushers came down Cunningham Gulch, after the struggle over Stony Pass, they could clearly see by looking left that King Solomon's mineral veins were standing out. Up some climbed and, on the North Star vein, a claim was located in 1872. Among others prospecting on the mountain came Winfield Scott Stratton whose 1875 claim did not "pan out." Stratton would have to wait for the 1890s Cripple Creek rush for fortune and fame.

The early history of the North Star (the highest claim on the mountain), like many of its neighbors, remains clouded. Until the Brunot Agreement of 1873, this would be Ute reservation. Not until 1874 did the territorial legislature create La Plata County. Two years later, San Juan County was carved out of the northern part of La Plata. Definitely by 1877, the mine was shipping ore that averaged a "little over 100 ounces of silver to the ton," but not a great amount (a total of ninety-nine tons) to Silverton and Lake City smelters. Lead and "some gold is mined as well." The mine's elevation, combined with winter weather, kept its operations seasonal. Still, development work continued. Silverton's *La Plata Miner*, January 18, 1879, in a long article described North Star development that included a sixty-foot adit with "mineral all the way," a twenty-foot tunnel, and a twenty-foot crosscut.

There were two North Stars in the Silverton area, the other being on Sultan Mountain within sight of the camp. At this time, the most famous North Star mine, however, operated in California, near Grass Valley and Nevada City, and the name was also found on mines elsewhere in the West.

Like so many early San Juan mines, the initial owners of the property did not have the finances to develop the North Star, and, as a result, leasers operated it in 1877-78 apparently with a profit for both parties. This changed when the Crooke Brothers, who operated the region's leading smelter over in Lake City, took an interest in the mine. New York capitalists and mining men, the Crookes—John (with twenty-five years of experience in "metallurgical operations"), Robert, Charles, and Lewis—built a concentrating works in 1876 that soon evolved into a smelter. In the following years, North Star ore was taken over the mountains to their works. Joined by several others in the vicinity, this allowed Lake City to revel in its fame as the region's smelting center.

The Crookes also became interested in mines, beginning in 1875 in the Summitville district in the southeastern corner of the San Juans. By 1879, it was estimated that they had "expended more than $400,000 on mines, works, and other property." To help provide ore for their smelting works and invest in promising mining property, the Crookes became involved in the North Star. As early as 1876, they acquired an interest in that mine. Once its value had been better established, they moved to consolidate their holdings, resulting, in 1879, in the incorporation of the North Star Mining and Smelting Company. Now with a New York company that had headquarters in that city and Lake City, Robert Crooke became their Colorado agent. The Crookes, like so many other eastern investors, dreamed of profits from the mineral rich Colorado mountains. Like so many others, they had lessons to learn.

By July 19, the *La Plata Miner* reported the mine was producing three tons of 150-ounce silver per day with a work force of eighteen miners. While the same paper reported (Nov. 29) the North Star paid "handsomely for four months," a month before the "beautiful snow" had caused the "boys to decamp in haste." Cheering for Silverton was the report that the Crookes intended to build a smelter at the mouth of Boulder Gulch. Every new mining district dreamed of having a smelter, or two or three, to cut transportation expenses and to treat those plentiful low grade ores that could not be shipped profitably. Quite often, sadly, those daydreams never became brick and stone.

The trials of early San Juan mining were shown when the company reorganized in early 1880, bringing in "some of the wealthiest men of New York state," with the Crooke brothers "still retaining their interest." The *La Plata Miner* (February 7,

1880) cheered the needed influx of capital, because "they mean business and know no such word as fail." Optimistically, they promised that the new smelting works would be up and running by August 1.

That projected mill and the North Star Mine were what brought the twenty-five-year-old New Yorker, Eben Erskine Olcott to the San Juans. Following "preparatory training" in the City College of New York, Olcott went on to Columbia School of Mines, graduating in 1874 with a degree in mining engineering. Journeying to North Carolina, he became a chemist in the Ore Knob Copper Company.

This was followed in 1875 by his becoming assistant superintendent of the Pennsylvania Lead Company in Mansfield, Pennsylvania. Like many others of his generation, he gained further experience by moving beyond continental North America. From 1876 into 1879, Olcott was in Venezuela working for the Orinoco Exploring and Mining Company, eventually becoming superintendent of their gold mines near El Callao. With this varied experience in copper, lead, and gold mining, he returned to the United States and soon journeyed West to report on mines.

It is at this point that Eben Olcott will tell his own story of his San Juan experiences and that of a few others that led to his arriving there. Despite his young age, he had gained a valuable working education over the past few years and acquired a better understanding of mining. This was no better shown than in a November 1878 letter to his sister Katharine:

> *"Sometimes I have wished that I had selected a different occupation, but then look what I would have missed. There are more wonderful satisfactions from being right than there ever are disappointments in making the wrong decision."*

## CHAPTER 1

# "I am willing to see the rough side of life"

$\mathscr{B}$efore going to the San Juans, Eben Olcott examined other mines in Colorado and elsewhere in the West during 1879-1880. His letters to his family give insights into the mining industry and its communities—and the people who lived and worked there during these boom years spurred on especially by the exciting silver discoveries at Leadville. He also makes his contact with the San Juans and the North Star Mine.

Olcott was a consulting mining engineer sent west to look at prospective properties for his employers. For example, he received a letter from D. Willis James of the Phelps, Dodge & Co. [1] on November 24, 1879 which advised him: "You know the properties in which I am *now* interested and in a general way after examination, I desire your opinion—whether they are *good or bad* and any suggestions, or advice that you have to give me." James then went on to list five Colorado and Nevada properties, including the Crooke Mining & Smelting Company.

Olcott and James had had previous discussions about Olcott's examining western properties. A letter of November 15 revealed the terms of the agreement and insights into the working life of a consulting mining engineer.

> *I will travel through the mines of the West for you on condition of your paying all my expenses to be limited to $10 per day & to be made as much under as possible and the sum of four hundred & fifty dollars ($450) per month during the time engaged in your service*
> *I would examine such mines as you designate sufficiently to prove whether they are safe business investments*

*and will endeavor to bring any others that offer particular inducements to your knowledge; either such as are already known to the public or encouraging prospects. In short to travel in your interests making all possible use of my own information and all that I can gather which from my acquaintances in the mining regions would likely be such as not usually brought to the public.*

Colorado became the center of his activities. The state had arrived at its second mining boom nearly a decade and a half since the Pike's Peak rush of 1859-61. Then came the bust of 1863-64. Now, suddenly investors raced to the centennial state, almost falling over themselves to invest, first at Leadville and then beyond. As author and artist Mary Hallock Foote wrote when she journeyed to the new silver queen, "all roads lead to Leadville." [2] Indeed, all roads led to Colorado during exhilarating 1879 and 1880 where silver fortunes seemed only over the next mountain.

Olcott's road also led elsewhere in the state other than Leadville, but the young mining engineer arrived in time to be in on the ground floor. He would visit both the older and the new districts to examine properties. Among those mining districts that blossomed because of the Leadville excitement was Silver Cliff in the Wet Mountain valley of Custer County. [3] It, and neighboring Westcliffe, looked upon themselves as another Leadville. So did many other hopeful communities, including several in the San Juans. Only Aspen, by the late 1880s, would equal the queen and proudly assume that title of Colorado's "silver queen."

On his way to Silver Cliff, Eben passed through Denver and Canon City. In a newsy December 7, 1879 letter written to his sister Pheme (as were most of his letters), he described the state's largest city and some of his plans. It also reveals much about the young man.

*Denver is a very pretty & active city—wide streets lighted with gas and a perfect net work of telegraphs &*

*telephones wires over head. I counted 80 on one pole on
one of the main streets. The plateau on which the city is
located is surrounded with mountains which now have
light snow on them. Some of the residences & villas in
& near the city are very handsome & the churches are
also reasonably fine buildings. The Cent. Pres. Church ⁴
where I was this A. M. is a pretty stone building & was
well filled. We had an exceedingly interesting sermon on
the resurrection and he [minister] advanced & maintained the idea that our bodies should also be raised up
at the last day. [Olcott was not sure about the theology
of the resurrection of bodies vs. souls and discussed his
belief.] However I say the sermon was very interesting &
good.*

    *I have made some very pleasant acquaintances &
seen a good deal. I leave for Boulder & Caribou ⁵ early
tomorrow A. M. to be gone 3 days. On my return I shall
either go to Leadville or Georgetown. You can continue
to forward letters here.*

    *The table here [Charpiot's Hotel and Restaurant ⁶]
is quite good though nothing astounding [about] the
rooms as these are for the most part poor, but I have a very
nice sunny one with a superb view. There is no furnace or
heater in the house and we have to have little stoves for
soft coal in every room. It has been so mild during the day
time there has been no necessity for fire until in the
evening. I have a little maid. The charges are certainly
high enough, 3.00 a day & fire extra.*

    *The Argo smelting works, Boston & Colorado
Co., ⁷ are very extensive & perfect. I spent all yesterday
morning there. That is the place where Grant when
President visited and they made a row of silver bricks for
him to walk on from the front door to the inner office
where he was received by Prof., now Senator Hill. ⁸ They*

*did not do the same for me. The sunset just now over the*
*distant mountains of the snowy range is beautiful.*

Olcott also spent a day visiting the coal fields near Canon City which he described as "all very good." Eben had received $100 to examine two properties there from a New York investor.

He found time to write Willis James four letters, December 8, 9, 18, and 31, showing that he intended on going to the San Juans ("It will take a week to reach & examine Crookes Smelting works"). He left the comforts of Denver and Silver Cliff, arrived there, and returned.

Olcott did explain the problems of traveling to the isolated San Juans as winter approached.

> . . . *at this season the San Juan region is so*
> *snowed up that very few of its mines can be seen & I*
> *probably would not see more than the smelting works &*
> *1 or 2 mines. I would not be able to get to Silverton &*
> *Howardsville at all. It will cost about $60 besides my*
> *time & it seems a good deal to spend.*
>
> *If you telegraph me not to go I shall get the most*
> *reliable information possible. I may have to pay a little*
> *for this but it will be less than my journey would cost*
> *and in the mean time will be busy at other things. I*
> *regret troubling you in this way but my instructions are*
> *so general that I find it necessary.*

James had already decided on the matter and in a telegram, dated December 9, advised, "Think you better go to the Crooke Smelting works."

On the same evening he sent the telegram from Silver Cliff, Eben discussed his interesting day examining mines and continued, "your telegram of today received and of course I will be pleased to go to Lake City to examine Crooks smelting works."

Olcott saw promise in the new Silver Cliff district, but also understood the fickleness of mining in this era. "I have had a very interesting day and have seen some mines which promise very well and as yet can be taken up very cheaply. Much depends upon whether the Silver Cliff [mine] makes a permanent success when her mill is erected." [9]

The December 18 letter was sent from Lake City, following his 125 mile "stage & sleigh ride" from Alamosa.

> *I have completed my examination of the property of the Crooke Mining & Smelting Co. both the mines & works and am very favorably impressed with the enterprise as a sure permanent investment. One of the mines the Ulay is excellent & will soon pay handsomely. The other the Ute is not sufficiently developed to pay. During the year of '79 I do not think there will be any profits, in '80 the profit must be at least $50,000, which will barely pay a moderate interest on the investment but in '81 & thereafter the Co. with good management will pay dividends of not less than 20% on amount of stock capital.*

Those enthusiastic comments certainly would not have dampened James' interest.

About the North Star, he went into some detail, with an excitement unusual for his letters and reports.

> *I desire to call your attention to the North Star Mining & Smelting Co. located on King Solomon Mt. near Silverton, San Juan. I have not seen the mine but have heard from it by several different reliable & disinterested authorities & seen the ore & studied the circumstances. It is 13,500 ft. high though unaccessible at present but well deserves an examination next spring. If you are [in contact?] with the New York management*

*office 7 Exchange Court N.Y., Pres. McPherson LeMoyne,*
*you could not get a surer investment than that stock at $5*
*per share par value $10 whole capital $250,000.*
    *This would be a permanent dividend paying mine*
*from the day smelting works were successfully operated.*
*For a year the stock would pay nothing. Mr. J. J. Crooke*
*who owns more than half the stock has made proposi-*
*tions to me to build smelting works there next spring &*
*summer and I would give positive assurance of success.*

Olcott went on to comment how the abortive attempts to build smelters in the San Juans had helped give the region a bad name. But there was more to it than that. "This and the kite-flying of worthless properties has hurt the region but mixed in with hundreds of mere prospects are some of the best mines in Colorado." Then he concluded, "I believe there is scarcely a better one all things considered than the North Star."

The proposed smelter held out great possibilities as well: "Smelting works doing custom work which the North Star works could do outside of its own ores would pay enormously at Silverton." Further, he saw a solution to the isolation and transportation woes of the region. Olcott had talked with David Dodge, Denver & Rio Grande vice president, [10] who "told me there could be a RR there by next August, even allow[ing] that it does not reach there till August 1881 now is the time to get" the works that "will treat the rich products of the San Juan."

Upon his return, Olcott sent James a long letter reiterating and expanding much of what he said previously. Of the Crooke company, he commented that the prospects included "silver-lead mines with various unexplored mining lands adjoining. Smelting works with necessary power & buildings, offices, etc, etc." Then he went into some detail on individual mines and the smelting process, but nothing more on the North Star in this letter. He examined the mines he could and other property carefully, even though "the snow was 4 & 5 feet deep when Mr. J. Crooke & I went up."

In that same letter, Olcott faced one of the major problems of the era, communication. "I cannot understand the delay as it [telegram] was promptly forwarded from Denver. I do not know what became of your telegram of 18th but in future I will make other arrangements with Charpiot." He then continued to discuss other properties but gently complained, "I cannot get through all examinations in the 3 months allotted to them if I stop too long in one place." [11]

Other investigations still needed to be concluded. Eben suggested to James that if he wanted him to return to Silver Cliff, to just jot "go back" in his next telegram, or to tell him to "go to Georgetown where I will look into the mines." To Silver Cliff he went and Olcott particularly recommended the Bassick mine at Rosita, near Silver Cliff. [12]

Olcott was delayed at Silver Cliff longer than was expected. "This is quite a geological knot to be unraveled." He wrote his sister on Christmas evening from the Powell House, corner of Main and DeWalt streets.

*Today the weather has been perfect 10 degrees below zero at breakfast time, but milder all day since with glorious sunlight and still. The moon tonight is glorious. Yesterday the mercury did not rise above 10 degrees below zero all day; during the night previous it was 20 degrees below with wind direct from the Rocky Mts. One certainly needs to be bundled up for such weather for it is terrific but if you manage to keep warm it is invigorating & splendid. Every morning I have to thaw out the water in my pitcher and it takes some moral courage to get out of bed.*

*There are nineteen peaks of the Sangre De Christo range in full view from here all snow covered, snow bon-*

*neted when there is wind, & cloud covered. It is a magnificent sight.*

*It is remarkable to see these new camps. This one built up in eighteen months now with hundreds of buildings well laid out; of course the buildings are mostly of wood for there is nothing like the capital here that there is in Leadville.*

*A holiday like this in a rough camp is sure to be the occasion of much drunkenness etc. but I expected to see rather more than I have. I guess however there will be quite a number to wake up tomorrow with a headache. There is no jail in town now as it was burned down last night and two poor wretches perished with it.*

*I am willing to see the rough side of life but certainly do not believe in the necessity of civilized Christian men falling into the ways of the majority.*

*I am very anxious to get to Leadville. At Leadville the smelting cos. are having great difficulty on account of the exorbitant price of coke and I hear on good authority that they are not making any money now. This will probably regulate itself next spring.* [13]

*This is one of the rising camps, very little silver has been shipped as yet but great things are expected for it next spring.*

On Sunday, December 28, Olcott again wrote his sister from Silver Cliff about this new camp and its people. For a man who had been engaged in mining for six years, his reactions are fascinating.

*I have enjoyed a very delightful Sunday. It was the day appointed for the dedication of the Pres. Church. This town of less than two years of age now numbers 4000 inhabitants large enough to support a church or two but when you take into consideration the class of people, rough & ungodly in the main, I think it is doing very well to have built a church and almost paid for it. This morning Rev. W. P. Teitsworth [14] preached & took up a collection towards defraying the balance due on the edifice, it was a very good positive pleasing sermon.*

*A half owner & tender in the bar of this the best hotel in the place, but very 2nd is named Mark Atkins. He is from Cincinnati and a well educated, very handsome fellow, and a Psi Phi. [15] A nice occupation isn't it!*

Eben did go onto Leadville, and he wrote from the Clarendon Hotel [16] on January 11, 1880 to Pheme, primarily talking about some of his South American acquaintances and adventures. He then went on to discuss some of his old friends that he had met in town and his injured knee. "I think my knee is decidedly better. I was much afraid that I had injured it seriously but now I think the swelling will be all gone shortly." How this accident happened, Olcott did not say. It probably occurred while he was climbing around underground.

Returning to Denver, Olcott found a letter for him from a contemporary young mining engineer John Farish. Farish, who would have a distinguished career, was at this time in Silver Cliff. After talking about some of the mines and mills there, he concluded, "I should like to keep posted as regards your movement & address as if I am successful in finding something that I can say is good, we may work together on a sale providing always that it suits you also." The mining fraternity was a close knit group. [17]

Meanwhile, Olcott received a flurry of letters about other mining properties from Farish illustrating some of the other activities of this young mining engineer. Some related to Silver Cliff, some to other parts of Colorado. A party from the former advised him that the Kate is on the market for $100,000 and advised him to "examine the property at once. Good properties are caught at as soon as in the market." By February 19, the Kate had been reduced to $75,000: "A bond for $75,000—60 days 10 per cent down and if you want an extension of 30 days, it can be had by paying 10 percent more at the expiration of 60 days." What was in it for Olcott? "If this sale is made we shall get 10% commission and will give you 1/3 of it," promised L. N. Clark, a Silver Cliff broker in real estate, mines, and mining stock.

Clark also told him that the Horn Silver was available for $50,000. Times would never be this exciting again in that district and town. Olcott also received two letters from Farish regarding owners, investors, and properties. From over at Georgetown, he received a communique about the Wisconsin Central Mines, whose owners were willing to sell four-fifths interest for $8,000: "Should you or your friends want this property you can have thirty (30) days refusal on the same." Older and known districts were not as pricey as newer ones! That fact of mining life was well understood. [18]

James was one of the parties interested in the Kate, but he did not invest. [19] On December 31, a disappointed Olcott wrote him, "Not having received any intimation from you that you were interested in the Kate I gave up all attempts at trying to secure it & have not examined it further. I still considered it one of the best prospects others holding the same opinion." All of his traveling around Colorado would help his investigations. He explained to James, "I have the advantage of acquaintances there & knowledge of the ground."

In early February, Eben was examining property near Frisco, in Beaver County, Utah. [20] In a February 8 letter, he described the problems of traveling in the West. The life of a mining engineer was not in the least romantic after a journey like this one.

> *When we reached the end of the R. R. last Thursday the only place to take supper was in the construction car with the workmen on the R.R. I had to wait till the 2nd table and then had to bolt my meal and hurry to take my seat in the miserable half open stage. There were four of us as passengers and not one slept a*

*wink between the discomfort of the cold, rough roads, wind, etc. Once the old stage almost tipped over and we all had to get out and help right it thereby covering ourselves with the dirt from the wheels, etc. The animals almost played out and took a distressing amount of laboring to make them go at all.*

*At about 5 A. M. Friday when we had all become so chilled that it was misery to continue we reached a dirty ranch where we were to have our breakfast. After awhile a fire was started which was very grateful and in about an hour a greasy breakfast was served. Then we continued on our way becoming 11 o'clock when we reached this place. There was nothing pleasing to the eye; everywhere was the sage brush & alkali plains with very little seen to hide the unsightliness.*

Olcott did not like much about the place or its inhabitants and had very little good to say about the local Indians or the Mormons. He described the latter in a February 19th letter, "the Mormons have made little impression on me. They are a bigoted race who spare no opportunity to injure a *gentile* but you may travel among them without noticing any marked features of their belief."

Commenting on Frisco, he continued, "There is no drinking water nearer than 15 miles from here and it costs when brought to town 4 cents per gallon. The washing water is brackish & so hard that your hands generally get an additional coating by trying to wash them." Nor did his accommodations bring forth praise. "The Southern Hotel is a half decent county tavern with the usual appendage of western roughs." Olcott concluded, "the mines I could speak more enthusiastically of than the general attractions(?)"

Trying to summarize his experience to Pheme, her brother admitted, "I don't know that it is worthwhile for me to complain for I have seen worse than all this in South America." Unsure of his plans for the future, he planned to travel to Galena, Lander County,

Nevada, [21] to look at mines there, but that fell through. In a February 19 letter from Salt Lake City, he worried about that "future." "I do not object to be[ing] busy, but I feel more restless to get at something permanent than ever I did in my life." There was a bright spot in all this, which Olcott realized. "This traveling is doing me lots of good. It is getting me known & I am seeing *worlds.*"

In the next month, Olcott apparently went home, as he had been planning to do for a long time, because, in a detailed letter to his mother from Palisade, Nevada, [22] he recounted his trip west from Omaha. "I find I have a long wait here so will recount my journey," and he went on to do that. Next on his agenda was to take the narrow gauge train to Eureka, Nevada. [23] Eben planned to leave there in a week for Colorado and gave his mother addresses to send letters to in Central City, Denver, and Leadville. Despite his concerns about Charpiot's Hotel, that continued to be his Denver address.

Back in Colorado he again wrote from Leadville on May 13. He enjoyed telling his family about his traveling experiences.

> *The railroad part of the journey up here was uneventful, we were detained seven hours. The Pullman sleepers on this narrow gauge South Park R. R. are the cutest little things you can imagine. They are furnished in exquisite-style, handsomer than any I have seen in the east. We saw a good deal of snow along the road and the 38 mile stage ride was nearly all accomplished in snow storms. This made the road bad. The coach ahead of us upset but fortunately no one was hurt to a greater extent than a few bruises.*

Olcott worried that people placed too much faith in finding silver fortunes at Leadville. He commented on it to his sister.

> *There was the gayest crowd you can imagine coming up. Eight R. R. cars loaded down with many of the passengers unable to get seats & there were twelve stage*

*coaches with from 12-15 persons each left Buena Vista*
*for Leadville alone to say nothing of the conveyances to*
*other parts, Gunnison, etc.* [24] *The thing is I think rather*
*over done, there will be many hundreds come back poorer*
*than they went away.*

Speaking of Leadville, he, like others noted, "Leadville is the same crowded wonderful camp with the roughs of all nationalities mixed in with the educated or refined all eagerly seeking to 'strike it rich' some where or some how." Of Colorado, Olcott observed, "Colorado certainly has the 'boom' this spring it will have the effect of bringing down labor to eastern rates as not one tenth part of the men who are here have means or strength to prospect." [25]

Eben felt Leadville was not a place for women. "I have met a host of acquaintances here, several people I know have their wives here but it is pretty hard times for a woman." Already he had made contact to work in the San Juans. "I hear from the place where my San Juan mine is that the large amount of snow there still blocks the roads; it will be a cheerful place to reside. Here it is the middle of May."

On his way from Leadville to the San Juans, he again passed through Canon City. [26] He had to take the stage, and the contrast with his Utah traveling experience was marked.

*As usual I was fortunate to get an outside seat*
*and despite the dust had a most gorgeous ride the scenery*
*being surpassingly beautiful. When we left it was quite*
*cold below freezing and in the cold morning my duster*
*was very comfortable and indeed I did not take it off all*
*day. We reached a place 61 miles from Leadville on the*
*Arkansas called Cleora* [27] *In the evening and I had*
*expected to take the train & get here about midnight but*
*we did not make it so came here this morning. It thereby*
*gave me the opportunity to see the gorgeous scenery of the*

*Grand Canon [28] by day and it is worth a trip to Colorado to see it.*

By late May, the San Juans were entering into his travels and, with another detour, also to Silver Cliff. Eben reassured his sister Pheme and their mother that he was in no danger from the Utes. "Practically speaking I have no more thoughts about Indians in Colo. than I have of Zulus. Colo. is a pretty big state and I have not been anywhere near Indian country and the reports of massacres have been very much exaggerated and in several instances notorious hoaxes for the sake of inducing the government to action in removing the poor red skins from Colo...." [29] Olcott had also caught the "fever": "Colorado is the place for me, such exquisite climate—intense cold in winter which you do not feel a suffer from & perfect springs, with the exception of wind; there are many wild flowers but not enough to make any great show as they are very much scattered but some of them are beautiful & would interest a botanist."

From Alamosa, on the twenty-fourth, he wrote his mother about his train ride from Denver. Unfortunately, the Denver & Rio Grande arrived late, and "the abominable coach has left and I will have to lose 24 hours."

*We had a very fine railroad trip by Denver & Rio Grande R. R. past the Spanish peaks, La Veta & [Sangre de Christo] mountains. The mule shoe curve which is the highest railroad on the continent except Meggs [?] Peruvian road and the Leadville line will soon exceed it in elevation. It is at any rate a splendid piece of engineering.*

*I think by another year [30] the R. R. will be completed to Silverton & then maybe I will ask the girls to spend the summer near there, but I doubt whether I get settled this summer and we will see how things turn out before another year.*

Olcott, who was the second oldest, had four sisters and three brothers. He worried in his letters about his family, bemoaned missing letters, and discussed various family matters and never lessened his ties.

Now, at long last, Eben was ready to settle in the San Juans where he heard Silverton "is quite a pretty place." The young mining engineer's days of traveling throughout the West, for the moment, came to an end. He had just turned twenty-six in March.

Eben Olcott during the past six months had journeyed through a significant portion of the 1879 mining West, especially Colorado, and to a lesser degree in Utah and Nevada. Along the way, he had gained further understanding and a practical education that would stand him in good stead in the San Juans. Isolation, inclement weather, poor transportation, high cost of living, life in the western mining communities, a sampling of the folk who drifted through the mining regions, and a taste of the mining conditions which soon he would confront—all had been observed and experienced.

For the educated, cultured New Yorker, this could not have been that strange and different. He possessed enough background in mining to grasp realities in most districts. However, this dedicated Christian found it hard, sometimes, to adjust to a community where, it seemed to many, that God had been "left at the Missouri River." The San Juans would present new challenges and some old familiar ones. From his letters, Eben looked forward to what lay ahead in those towering mountains. For a fellow who enjoyed spectacular scenery and snappy winter weather, he was about to enter a wonderful environment. Like many of late nineteenth century contemporaries, Eben Olcott would be in awe of what he would see in these high and rugged mountains.

1. D. Willis James was the son of one of the founders (Daniel James) of the Phelps, Dodge & Company, he played an active role in the company in the 1880s. Interestingly, 1880-81 was when Phelps Dodge sent James Douglas to examine copper mines in Bisbee and the Morenci, Arizona—they became the pillars of the company's copper fortune.

2. Artist and writer, Mary Hallock Foote went to Leadville in 1879 to be with her mining husband Arthur. See Rodman Paul, *A Victorian Gentlewoman in the Far West* (San Marino: The Huntington Library, 1972).

3. Prospectors had been poking around the district for most of the 1870s before Silver Cliff boomed in 1879 when Olcott examined there. The next year it had a population of 5,040 which declined abruptly to 546 in the 1890 census.

4. This is not the current, well-known Central Presbyterian Church at 1660 Sherman. The congregation had been started in 1860 and the church Olcott attended was a fore-runner of the present structure.

5. Boulder had been the site of one of the three original gold discoveries in 1858. Caribou, located twenty-two miles west of Boulder near the Continental Divide, had boomed in the first half of the 1870s. It was Colorado's northern most silver district.

6. Located on Larimer Street, Charpiot's advertised itself as the "Delmonico of the West." Despite Olcott's comments, it was one of Denver's finest hotels, "the place to stay," and its eloquent restaurant equally famous.

7. Colorado's leading smelter at this time. Located north of Denver and owned by Nathaniel Hill, it had opened on January 1, 1879.

8. Olcott has his story mixed up. The silver bricks were laid in front of Central City's Teller House in 1873 for President Grant's visit. They came from Caribou. Nathaniel Hill was the man who solved Colorado's refractory ore problems and his smelter, before moving to Argo on Denver's outskirts, had been located at Black Hawk. He served one term in the United States senate, 1879-85.

9. Olcott was right, the Silver Cliff mine and mill were key to 1879-80. New York investors purchased the mine and organized the Silver Cliff Company in 1879. Its mill went into operation in January 1880, but faced various delays. Frank Fossett, *Colorado* (New York: C. G. Grawford, 1880), 472-73.

10. Dodge, a long time employee of the D&RG, was involved in a variety of activities including reporting on Leadville as a possible destination point. As late as the nineteen teens, he was involved in Colorado railroad matters

11. All of these letters and telegrams to and from James are found in the archives of the New-York Historical Society.

12. Rosita, founded in 1872 and once the Custer County seat, had a population of 1,008 in 1880. Like the rest of the district, it declined thereafter. Located seven miles from Silver Cliff, Rosita lost the county seat designation to that community, which in turn eventually lost it to Westcliffe. The Bassick mine was one of the most promising in the district. See Fossett, *Colorado*, 467-70.

13. Leadville, because of its wealth, soon resolved most of its problems (including having fourteen smelters operating in 1881), but the coming of the railroads in 1880 helped most of all. Its population in 1880 reached 14,820, second largest in Colorado.

14. Reportedly, Rev. Teitsworth, serving at Rosita at the time, saw a need and preached the first sermon on the streets of Silver Cliff. The Presbyterians held the initial church serv-

ice and were soon joined by Methodist, Catholic, Baptist and Episcopal churches. See Muriel S. Wolle, *Stampede to Timberline* (Chicago: Sage Books, 1974), 291-92.

15. Unless Psi Phi was a local fraternity, Olcott has his Greek letters mixed up. The only fraternity in the 1870s with those two letters was Phi Kappa Psi. Columbia had a Phi Psi chapter dating from 1872. John Robson (ed.), *Baird's Manual of American College Fraternities* (Menasha, Wisconsin: George Banta Company, 1963).

16. The Clarendon was Leadville's finest hotel at this time. Built in 1879, it sat next to the Tabor Opera House on Harrison Avenue. Its restaurant was celebrated for its food, the chef being from Delmonico's in New York.

17. Parish to Olcott, January 20, 1880. This letter and those of James to Olcott, Nov. 24, 1879, and Olcott to James, December 18 & 31, 1879 are found in the New-York Historical Society. All other letters and material, unless otherwise identified, are from the American Heritage Center, University of Wyoming.

18. W. Francis to Olcott, Jan 7, 1880; Clark & Wills to Olcott, January 8, 14, February 10 & 19, 1880; Farish to Olcott, January 9 & 13, 1880; and Richard (?) & Pomeroy, January 20, 1880,.New-York Historical Society.

19. Perhaps that was a wise decision. Frank Fossett in his *Colorado*, published in 1880, does not mention the Kate among the several scores of promising prospects he lists. John Farish liked the property also, as did several Silver Cliff correspondents, who may have been financially interested in it.

20. Beaver County was famous for its silver mines, with the first mining done in the 1850s. The great rush came in the mid-1870s, with the discovery of the Horn Silver Mine. It was a prospering district when Olcott visited there. In 1882, diamonds were reportedly found twenty miles from the camp (*Engineering and Mining Journal*, November 11, 1882). Frisco, now a ghost town, was located in the north central part of the county.

21. Silver was discovered in 1863, but what small boom it had did not reach Galena for another decade. By the time Olcott planned to go there, it had slipped into the downward cycle.

22. Founded as a station on the Central Pacific Railroad in 1868, Palisade served as the "jumping off" point for nearby mining districts, including Eureka and Mineral Hill. Olcott was here to board the Eureka & Palisade Railroad.

23. One of Nevada's premier mining towns, Eureka was in full stride in the 1870s. Olcott arrived at the very peak, in 1878-79. Lead-silver smelting was pioneered in this district, and the methods spread to Colorado and elsewhere. The West's first significant apex case was argued in Eureka in 1877-81, between the district's two major mining companies fighting over a rich silver deposit. This brooded ill for the mining West because who controlled the apex (the top of the vein) controlled the vein downward through the side limits of the claim. The uncertain nature of geology, at this time, led to costly, savage litigation to prove who had the apex.

24. The Gunnison country was one of those districts that gained new life from the Leadville excitement. Its "excitement" was in full swing in 1880.

25. While in Leadville on this trip, Olcott took a ride up California Gulch where the gold discovery had been made in 1860. During the ride, he "made the acquaintance of one of the earliest discoverers of gold here before silver was known to be present, Mr. Stevens a rough looking specimen but worth a million." This was probably William Stevens, who helped open the silver deposits in the second half of the 1870s.

SAN JUAN GOLD: *A Mining Engineer's Adventures, 1879-1881*

26. A jumping off place for several mining rushes. including the 1861 and 1870-71 ones to the San Juans, Canon City was also near coal fields, as Olcott mentioned, and Colorado's first oil wells. The Denver & Rio Grande went through here on its way to Leadville.

27. Olcott reached Cleora just in time for its only "boom." Named for the daughter of a local family, it had been started by the Santa Fe Railroad during its fight with the D&RG over Royal Gorge. With the victory of the latter, nearby Salida, a D&RG town, replaced Cleora which quickly declined to virtually nothing.

28. Royal Gorge, one of the scenic attractions on the line of the Denver & Rio Grande Leadville route. In the eyes of some Coloradans, it rivaled the Grand Canyon. Olcott was on his way to Silver Cliff and Denver again.

29. He is referring here to the trouble with the White River Utes in September and October 1879, which resulted in the death of agent Nathan Meeker and eleven men at the agency, near present Meeker and a brief "Ute War." Olcott was right, this did lead to the call to remove the Utes, and all were gone by the end of 1881, except the Southern Utes on their reservation near Durango.

30. Olcott was a bit optimistic about the railroad. It did not reach Durango, a town the D&RG started, until July 1881 and went on to Silverton a year later. See chapter 3.

# CHAPTER 2

## "I am ordered
into action"

$\mathcal{T}$he potential of the North Star definitely intrigued Eben Olcott. Although he had not been able to personally examine it, in April 1880 he signed an agreement to purchase 8,330 shares for $25,000 in the company, "subject to the conditions of syndicate [North Star] agreement with the Crooke Brothers." Those conditions were unstated, and, perhaps, he purchased stock for D. Willis James or some other investor. Prudently, he had until August 1 to decide. It seems doubtful, this early in his career, that Olcott had that kind of money.

He had become fascinated with the North Star's possibilities from what he had learned and from his discussions with John Crooke. Eben moved to the San Juans and took on permanent work, although, at the moment, without an employment contract. His consulting engineering business was not placed completely on hold, as will be seen in several of the following letters.

The agreement Eben eventually signed on August 24, 1880 with the Crookes was probably the best the young mining engineer had yet seen. Hired as agent and superintendent of the North Star Mining and Smelting Company, he would be paid $500 per month, "payable at the end of each and every month together with his living expenses, keep of his horse needed in the service" of the Crookes, and "traveling expenses." In return, he not only would be involved in the North Star operations, but, when not needed there, "will assist the operations of the [Crookes] in any way in his power in the life of his profession as Mining and Metallurgical Engineer as such of their properties as they may direct." The agreement would run through January 1, 1882. Eben had already been on the job several months before the two parties finally agreed to the contract. The Crookes could be sometimes slow like this, or perhaps Eben wanted to know more about what he was getting into in the San Juans. [1]

His remuneration was better than average for superintendents in this era, although the basic salary varied greatly depending on the size of the property and the experience and reputation of the individual. Some received as low as $250 per month, others $400, and a very few, at the top of the profession, nearly $1,000. Occasionally, they also received stock and sometimes a percentage of the mine's net profit. The horse and expenses were not unusual guarantees, and sometimes the company provided a house. Risk existed in all this. Unlike his miners, a manager was not entitled to a lien on the property to secure his unpaid salary. If the company went bankrupt, he might have great difficulty collecting his back wages. [2]

Olcott's first letters after he arrived in the San Juans came from Lake City. Initially, he stayed in the American House, [3] but he quickly found a boarding house. Once again the perils of transportation graphically came into view in his May 27 and 30, 1880 letters. Eben, too, found out about cooking at higher elevations. These also were addressed to Pheme unless otherwise noted.

> *[May 27, 1880]*
> *Only a few words tonight. I have had a very slow trip down here as you have doubtless learned. Today we had an adventure with a broken coach-wagon which threw three of us from the open wagon without scratching or hurting anyone of us at all. It was a very fortunate escape from what might have been an ugly accident. I was thrown under the pole of the wagon and my face touched the horses legs. The driver landed on top of me and we managed to hang in some way so as not to go to the forward. The cause was an axle in the front of the wagon breaking as we were going downhill. The ride was glorious as scary — we had snow today and snow*

*several feet deep in the woods. I had to walk today a long
way as the wagon would not accommodate all the passengers. I have walked at least 11 miles some of it over a
mountain 11,500 feet high. I am not tired one iota.*

*I do not know about my employment yet; have had
a lazy week with J.J.C. I am ordered to wait at their
expense ($500 per month) for the present perhaps a
month. I can't object though. I would much rather be
working than idling but I guess I will have something to
do for them. I may not be home till Aug. I received a
telegram from Howell today that they expect to purchase
the Silver Conisor [?] are pleased with my report and
would like to get me for supt. I am very glad but expect
to make a permanent thing here.*

*I received today on reaching here an* Engineering
& Mining Journal [4] . . . *No letters as yet.*

[*May 30, 1880*]
*I have first finished a volume to Frank and have
a little time left for a home letter. I feel as though I was
quite established. I have a comfortable little room in a
boarding house and a P.O. box. The latter I thought was
worth $1.00 for 3 month as it insures my getting my
mail much better and is more convenient.* [5]

*You may be interested in hearing more of my stage-
ride from Alamosa here. It is in all 112 miles about 87
miles of which is along the valley of the Rio Grande. In
places it is wide and affords opportunity for farming by
aid of irrigation without which there is no cultivation, as
there is such a lack of rain. After you pass Del Norte* [6]
*however the river is confined mostly within the rocky
canon and the diversity of eruptive rocks then present
themselves is wonderful. In some places the road is dangerous but is generally fairly well constructed. At Wagon
Wheel Gap* [7] *is the most peculiar and noted scenery on*

*this route. The red volcanic rock stands up almost per-
pendicular on either side of the river and now the Rio
Grande comes down a foaming current. Soon after leaving Antelope Springs <sup>8</sup> too we pass, to my mind, the most
picturesque scenery on the road. Just before coming to
Lake Santa Maria, there we saw very many wild ducks
and geese and at two places we saw antelope. The first
night I stopped at Del Norte and the second at Clear
Creek Station; at Del Norte I slept in a room with three
other men and at Clear Creek with six others but by a little stratagem I secured a bed alone the second night. Its
tumble in anyhow in this country. I have most remarkable fortune in always getting the pick seating though I
want the upper stage seat and I scarcely ever sit anywhere
else and have only been forced inside for a ride of 15 miles
in all these hundreds of miles staging.*

*The night we were at Clear Creek it was very cold
even an inch of ice forming near the house; from there
the stage company only ran a spring wagon which is
supposed to only accommodate five besides the driver,
mail and luggage but seven insisted upon going in it
and I volunteered to walk a good deal. We reached the
Continental Divide with no other incident than a war of
words waged between the driver and a couple of passengers but soon after we started to go down the hill, I who
was sitting in the dash board was precipitated into the
pole and no sooner than I comfortably landed when this
driver came on top of me and I saw the other man who
occupied the front seat fly off at a tangent in the direction of the road a few feet ahead. How I managed to
cling on to the pole and finally get my hand back to the
dash board and the driver to hang on to the breeching of
the horses and both keep from going on to the ground is
a mystery and was most merciful and providential for
had we gone to the ground we would almost certainly*

*been run over and likely killed. My face in falling
touched the back of one of his horses's legs. If it had
kicked I could scarcely have escaped. As it was not one
of us was hurt or scratched. The front axle had broken
and thrown everything forward. I mounted one of the
horses bareback and rode to the nearest station where
provision was made for carrying on the passengers and
I started in on foot over the highest part of the route. In
fact I believe the summit of the road is the highest place
I have been to; it is 11,590 feet above sea level. It is
funny to see the face a cook makes if you order boiled eggs
which we at home think are almost the easiest thing to
prepare but out here they take 15 min. to become slightly
congelation. You see water boils at such a low temperature owing to this rare atmosphere that there is not suf-
ficient warmth generated. At Leadville you can put your
hand in boiling water and at the North Star Mine it is
said boiling water will not scald the face. The temperature of boiling water at Leadville is only 90 degrees Fah.
(Chester says this must be 190 degrees Fah or else 90
degrees centigrade). Some friends of mine who are batch-
ing complain bitterly that they never get time to boil potatoes (3 hours) always having to eat them fried. Beans
take an age to boil and even rice is very apt to be half
hard when served. ⁹*

*I felt no unpleasant effect from the rareness of the
atmosphere at over 11,000 feet ¹⁰ and I was walking
briskly. Of course I was conscious that I had to breathe
fast but it is rather pleasant than otherwise. In drawing
on a woolen shirt it snaps and crackles like wood burn-
ing. So to brush your coat and you feel this electricity tingle in your fingers. While my gutta percha drawing
triangle is moving over the paper it gets covered with dust
and fragments paper to such an extent as to become a*

❧ 33 ❧

*nuisance. Altogether the day I came in I walked ten or eleven miles.*

*One of the very highest mountains in America is Uncompahgre 14,484 ft.* [11] *That is very near us here frowning down with generally a storm brooding about it and I have several times been in the clouds.*

*Two streams unite by the town here. Hensen Creek and Lake Fork, the latter forms very beautiful cascades which are made to do service to Crooke as a water power. The Presbyterian Church* [12] *here is temporarily closed and the morning I attended the Episcopalian. After service the rector said to me that he was glad I knew the service so well. These seemed to be only two others who followed at all. It was a rather feeble attempt all through but the rector was exceedingly pleasant and good. The people living here are not at all as familiar with the forms of that church and in fact there is none too much church going of any kind. The edifice is a little bit of a rough building that looks like a carpenter shop but inside is very neat.*

*There goes the church bell for evening service so I will close.*

Sunday being a wide-open day in western mining communities, the only day in fact many miners did not work, churches had to compete with everything for a miner's attention from business to the red-light district to sports. Women and children predominated in the congregations. The church did offer the best family social outlet available and an organization where women could provide leadership. Churches often spearheaded the drive for Sunday closing of saloons, and they emerged a bastion for the WCTU and its activities.

On June 6, Eben wrote "My Dear Mother." While the letter mostly pertained to family matters, he gave a few insights into his San Juan experience, including one interesting fellow, the type of individual that occasionally appeared in the mining west.

*This letter must be short tonight and indeed I have written several times this week so that there is little to say. My friend Count (not Chevalier as I wrote) du Payer was ill last night and I want to call and see him before I go to church. I thoroughly enjoyed the dinner I gave him and friend. The old landlady got up a most creditable meal - Julian soup, roast veal, Boston browned potatoes and tomatoes, asparagus, plum pie, coconut cake and coffee and the fellows enjoyed it I think. du Payer is a remarkable genius graduated at the Ecole de Niue de Paris and that is an honor and he has practiced in Ecuador, Nicaragua, Honduras, Venezuela, Brazil, Italy, the U.S. and elsewhere and is a first rate engineer accustomed to the best society in Paris and Hungary and yet content to drive a burro laden with the few worldly possessions he is now possessed of.*

*This morning the Episcopal service was so pleasant. They were nine people there and both the choir and minister were more pleasing than last week.*

*Last night I fixed two silver bricks for mailing and they may go by the mail. One is for Aura and the other for Emma. They are miniatures of the ordinary 100 lb. bricks that are shipped from here and most silver works. They are fine bullion containing 998 parts in 1000 of silver and gold (mostly silver), and are just as are sent to the U.S. Assay office in N.Y. and Phila. for parting where the silver is dissolved out and the gold filtered off. They are the same shape and proportion as the big ones*

*made by the assayer and me yesterday in Crooke's labo-*
*ratory are in fact miniature facsimiles of the real bricks.*

Three days later, he wrote Pheme about a planned trip to the
North Star and then on the 16th a longer letter from the mine.

*I am ordered to action. I leave here on Friday*
*morning (11th) for a 30-mile mountain climb to exam-*
*ine the North Star Mine and while there will have few*
*opportunities of writing. I expect to be back here in 10 or*
*12 days but may be longer. I have orders here for*
*telegrams to be dispatched by special messenger if any*
*important ones come. There is no station at Silverton.* [13]
*You may not receive any letters till I get back so you need*
*not be alarmed.*

*There are three of us going on the expedition on*
*horse back. It will seem like S.A. all except the snow. We*
*still have to select our route not going the shortest way in*
*consequence of snow, etc.*

Olcott appeared to become frustrated that he had not been
more actively involved with the mine or proposed smelter. The
projected smelter, at the mouth of Boulder Gulch to process
North Star ore, seemingly had been placed on hold. He had discussed this with the Crookes back in December, while examining their property, and he seemed excited about the idea. They
then had planned to build the works in the coming spring and
summer to place them nearer the mine thereby reducing freight-
ing expenses. Eben, as many others had earlier, had found the
Crookes to be unpredictable in what they might—or might
not—do. They had the best intentions, but its execution sometimes failed to match. The planned Boulder Gulch location

would have been a little more than three miles northwest from the mine, down along the Animas River.

Those problems did not undermine his appreciation for what he was viewing as he traveled about the San Juans. He wrote Pheme:

*[June 16, 1880]*

*I am just through my work and as I have to go to Silverton tomorrow A.M. want to write you a little note. I cannot do much.*

*I am reveling in wonders. We overlook hundreds of Rocky Mt. peaks. Tonight I counted 20 in one direction that I could see without turning my eyes at all. The sun rises and sets in a glorious sea of colors and the snow and frost effects are most entertaining. The clouds too are wonderful.*

*The trail up here ascends 4,200 feet in not over 3 miles and lies along the edge of precipices but it is not as dangerous as I supposed it was from accounts — in fact I was disappointed by its tameness. We can look down however nearly a thousand feet almost beneath you and if the animals should fall it would be fatal. On the steepest places we walk and take hold of our horses tails to assist us in summiting the hills. The little animals go along bravely and it is ridiculous to see a string of men in this position. This extreme altitude is peculiar but not at all distressing and I am enjoying the experience very much.*

*I have been here now 3 nights and days. I shall remain a week more then ride down to Silverton only takes 1-1/2 hrs but it takes 3-1/2 hrs. to come up.*

*We have quite a number of wild flowers way up here while timber line is 1,500 feet lower down. I enclose two or three little blossoms picked on Engineer Mt. 12,600 feet high [14] and have one or two from that altitude. People say that the profusion is wonderful during*

*the summer and you understand we are fast getting*
*spring now. I can stand at my room don't [have to]*
*throw a stone long to snow some eight feet deep and I*
*know ravines where there is 100 feet of snow which never*
*melts away. It commences to freeze soon after 4 p.m.*
*every day and the nights especially from 2 to 3 a.m. are*
*very cold.*

*We are living rather simply. I have eight miners*
*here and Capt. Cruise is with me. We found more work*
*to be done than we expected to. Consequently are rushed.*
*I do not expect to make a permanent location here. I do*
*not think I could accept the position if they wanted me.*
*To tell the truth* che nos *I am disappointed in the mine*
*especially the development. It looks as though I should be*
*busy till Aug. 1st at least*

Olcott might have been disappointed then, but, as indicated
earlier in the chapter, he signed a contract to stay until 1882.
Reaching the mine for the first time, and realizing the transportation and weather conditions, no doubt accounts for a large portion
of his sudden change of mind. New Yorkers were not used to seeing eight feet of snow outside their door in June, nor even, apparently, a veteran North and South American miner.

Olcott was not exaggerating about the trail into the mine,
which perched upon one of the most challenging mineral loca-
tions in the San Juans. The mine was "touted with some degree
of accuracy during the 19th century as the highest consistent
producing mine on the North American continent." [15]

The next letters from Lake City and Denver deal with a
variety of matters, often discussing home and family. Eben was
doing a lot of traveling around the San Juans, which he often
recounts in some detail. The Crookes appear to have him running errands, though perhaps some of them were his own activities. Meanwhile, operations at the North Star resumed, despite
the disappointment that the property was not in the condition

he hoped, or that in the condition that the Crookes had led him to believe to be the situation on King Solomon Mountain.

*[June 26, 1880]*

*My dear Pheme,*

*I reached here yesterday after a very pleasant ride over from Silverton, as usual I found one of your interesting letters which I immediately devoured, I think almost reading it in the Crooke's face before I answered his thousand and one questions.*

*I expect to leave here on Tuesday (29th) for Denver where I will meet the party who came to the North Star.* [16] *We may make a negotiation out of that mine after all. I have bought for my use this summer the prettiest pony you can imagine and a perfect gazelle in climbing steep mountain trails.*

*Today I have been very busy examining the specimens I brought over from the mine. I will have to try and crowd two days work into Monday for I leave early Tuesday morning for twenty four hours staging. I would ride my horse out to the railroad if I had time.*

*I believe I did not write that last Sunday night. I had a slight stroke of lightning in that magnificent thunder storm we were in for several hours. It was not more than a good shock from your electric battery merely making my head tingle for a second or two.* [17]

*I wish I could have pulled a few of the many wild flowers I saw, some were beautiful and entirely new to me with the exception of three or four that I recognized as garden flowers at home, such as the musk plant, rocket flower and a little blue bell.* [18] *The girls would go wild over the woods here. I hope you will see them some day.*

[June 28th, 1880]

My dear Mother

*I have passed a very pleasant Sunday. Two very good sermons morning and evening and a most beautiful walk this afternoon up to Lake Fork on the Gunnison. There are two falls in this river within four and a half miles of home. The second one is magnificently beautiful. It dashes down about 80 feet throwing up immense clouds of spray which envelope me in a most refreshing atmosphere on a hot day and the sun in striking it makes a beauteous rainbow. Only half a mile above this fall is a lake four miles long by half a mile wide which is very picturesque. The day was beautiful and the numerous wild flowers added much to the attractiveness of the woods.*

*The girls must someday see Colorado, when I get a house of my own. I meant to write a long letter today but now it is late and I have a good deal to do tomorrow and must go to bed. I will likely be in Denver for the celebration of the Fourth of July. God knows I have missed this day for four years. Not that I expect to take any childish delight in it.*

*It takes about a dozen of my notes to make one respectable letter.*

[Denver, Colorado]

[July 5th, 1880]

My dear Mother

*The gentleman I am waiting for might soon be at his hotel and I shall try and get him to start on his journey tomorrow but being elderly he may desire a day's rest after three days of travel from San Francisco. As I did not have to be here till tonight I decided to run up the country and examine a mine. I returned today and make a thorough examination on Saturday and shall make a little extra money thereby. One of the gentlemen for whom I made the examination is James P. Lowry. One of our party back to Silverton will be Mr. E. R. Bacon of the firm of Field Dorsheimer Bacon & Dego (Dudley Field LC). I have seen a good deal of him since he has been here. He is pleasant but like a good many young lawyers puts on a good many airs.* [19]

*Denver is very pretty now in its summer foliage and the distant mountains showing palettes of snow look beautiful. This last little trip of mine was through magnificent country for scenery but I can assure you I had to rough it at the mine; slept in a cabin that was not fit for respectable cattle. It was almost as high as the North Star Mine and to climb to it was worse. I shall send the report to Pheme to copy before long.* [20] *I wish you would see whether she is provided with my report papers so that in case she is not it can be sent when my report arrives.*

*The city is ninety today and very hot. It quite seems as though I might be in the neighborhood of New York.*

*I am tired and not at all in a letter writing mood.*

*[July 17, '80]*

My dear Pheme

*I have been ordered home again on business. I go today to Del Norte to examine some property near there and then I expect to start home and calculate to reach there about the end of the month. I sent a night message to Jake informing him and telling him not to forward any more letters as I will not be back here. I presume you* will inform Emma on such occasion when I telegraph my movements so that she will stop sending letters.

I have a little more work I meant to do out here before I returned but something has come up which Crooke wants me to attend to in New York.

It is about six o'clock and I have been up some time and had my breakfast and am now some little ahead of the stage.

We had some ridiculous experiences with Bacon, Dudley Field's partner. He was not used to traveling on horse back in the first place, then he lost one thing after another and wound up by losing a gold watch chain trinket he valued at $5,000.00. After accusing three men, one of which he was sure had it, and raising a hubbub here, I received a note from up country saying one of the party left a watch under his pillow ten miles from here. Then the coward instead of going to see the men he accused ran off yesterday morning immediately upon receiving his property and let a lawyer inform the men or rather wrote a note to his lawyer who was so mad that he would not have anything to do with the matter. If Bacon had been here another night he would have been horse whipped.

I am afraid I will not get another letter before I leave but it is possible I may. My last is of the 5th. You were very good to write — both on the 4th and 5th. The second was certainly a very pleasant surprise. I don't think I will stop at all coming home though I may remain one night in St. Louis.

*I have been very busy lately and have had no time
for writing and may not be able to write again before I
reach home. If I change my plans I will telegraph.*

*I have no time for more.*

*[July 20, 1880]*
*Little Annie Gold-Mining Company* [21]
*SUMMIT*
*Rio Grande County*
*Colorado*
*C.E. Robins, - - - - Financial Agent* [22]
*Dear Pheme*

*This will be the last letter I will write before I
return. My plan is to arrive home about the 30th. I have
had quite a pleasant time up here tramping over the
mountains and examining the underground work on
this mine. It is a perfect stickler and I am having hard
work to make order out of the chaos of veins and
deposits. I do not think I might stop anywhere on my
way home but very likely I will remain over one train at
St. Louis as I want to see a man on business. I am having a great experience this year above timber line. This
makes the fourth place I have been 12,000 feet or more
high.* [23] *I can't say I prefer this elevation to one more
moderate. I don't think it is particularly healthy as a
permanency but for a little summer experience it is well
enough. Miss Robins returned today.* [24] *She seems to be
quite bright and not afraid of hearing herself talk. Her
father has taught her to assay and is putting in practice*

*my theory of the opening in this field for women. The Robins are from Staten Island.*

*I will let you know from N.Y. my plans. I cannot say anything about them till I have seen some parties in N.Y.*

*I have got a lot to see yet of western things and ways. Every time I look over these mountains I think how you would enjoy them.*

What Olcott might have thought of the Little Annie he did not say in a remaining letter. That mine had been noted for its rich specimens of gold, including one that ran $160,000 to the ton, which had helped stir interest in the San Juans back in the early 1870s. It also ranked as the most developed mine in the district with a tramway, mill, bunkhouse, assay office and other buildings.

The district soon peaked and declined for the first, but not last time. The report of the United States Commissioner of Mines for 1882 noted simply, "the Little Annie was also an early discovered mine, and for a time was considered to be the most valuable in the district." By the next year only one mine, the Golconda, was producing.

At this point, Olcott returned east, and his letters stopped for the next month. He did not elaborate about what business the Crookes wanted him to attend to, but, during this time he signed his employment contract. He would return to the North Star with renewed enthusiasm and be as romantic about the environment as ever.

1. This agreement and the stock transaction were found in the Olcott papers, New-York Historical Society.

2. Clark Spence, *Mining Engineers & The American West* (New Haven: Yale, 1970), 192-94.

3. Proprietor William Connor promised "no pains will be spared to make its patrons comfortable." It had recently been remodeled and enlarged and made into a first-class house.

4. Published in New York, the *Engineering and Mining Journal* was one of the leading American mining journals. Throughout his early letters, Olcott commented about receiving, or not receiving, the journal.

5. Here Olcott discusses some photographs. Family matters have generally been edited out, unless they have some direct bearing on Olcott's mining career.

6. Settlers originally reached the area of Del Norte in the 1860s. The town served as one of the gateways to the San Juans and came into prominence in the 1870s. Stony Pass led over the Continental Divide into Howardsville and Silverton, and Spring Creek Pass and Slumgullion Pass to Lake City. Olcott crossed over the last two passes at 10,091 and 11,361 feet.

7. Named for some wheels of unknown origin, most likely from the 1860-61 Baker excitement in the San Juans, a tiny settlement here actually had a post office when Olcott passed through.

8. Antelope Springs was on the Rio Grande route to both Silverton and Lake City. It contained a post office, store, and "public house." Ranching was carried on in the vicinity.

9. High altitude cooking received many comments in diaries and letters throughout these years. Visitors were amazed at the temperature water boiled. See, for example, Harriet Backus, *Tomboy Bride*.

10. Potential altitude problems worried early Coloradans and visitors alike, and some did come down with "altitude sickness." Conversely, Colorado, because of its higher elevation, contained more "ozone," which was believed to be a great health benefit, especially for people with lung and respiratory problems. Their state was therefore hailed as a wonderful health resort.

11. Actually, Olcott was not far off. The peak is 14,309 feet. The first recorded ascent was in 1874, by members of the Hayden party. The peak was named after the river and is a Ute name.

12. The Presbyterian Church was started by the famous pioneering minister George Darley and his brother Alexander in 1876. It still stands. No minister worked harder in the San Juans in the 1870s than George Darley.

13. Silverton's telegraph line, when completed, ran to Durango and beyond, not over the mountains to Lake City.

14. Around Engineer Mountain (13,218 feet) curved one of the two passes from Lake City into the Animas River watershed.

15. Allen Nossaman, *Many More Mountains,* (Denver: Sundance Publications, 1998), vol. III, 21.

16. The Crookes seemed to be willing to sell the North Star, if the price was right. Whether or not James might have been involved in the interested party is not clear. For whatever reason, the deal fell through, and the Crookes launched greater mining efforts than before.

17. Olcott was lucky here. More people were killed in the San Juans in natural accidents—snow slides, floods, lightning, cold weather—than in mining.

18. Blue bells are common to the San Juans. The musk plant may be what is locally called the "skunk cabbage." An *1888 Burpee's Farm Annual* identified one of its Larkspur varieties as "rocket." Larkspurs do grow in the Rocky Mountains.

19. Lawyers were not always "beloved" in the mining West. Westerners often objected to their "taking" property from hard-working miners by "legal" means. Some mining districts actually had laws against them practicing, and a few went so far as to forbid them

from even owning a claim. Mining disputes, however, proved a fertile ground for lawyers, and they practiced everywhere.

20. This explains why Pheme received the vast majority of the letters that still remain. She did secretarial work for her brother. In another letter (July 17), Olcott advised her, "Please collect enclosed as receipt. I have charged it in account for copying."

21. The Little Annie mine was owned by the Crooke brothers, and they put up a ten-stamp mill on the property. It had been one of the early important discoveries in the Summit district, the first of the San Juan bonanza districts. Gold was the principal mineral. Thomas Bowen, the earliest San Juan "millionaire" and later United States senator, made his fortune here. The Crookes also were partners in the neighboring Golden Queen mine.

22. Fossett's *Colorado* (1880) lists Robins as treasurer of the Little Annie Company. The mine from 1875 into 1878, yielded $164,000 in gold. His daughter must have been one of the earliest, if not the first, woman assayer in Colorado.

23. Actually not quite that elevated, some points did top 11, 000 feet however and the North Star 12,000. The isolated camp of Summitville was located in this district (one of the lower ones in the San Juans) and "reached by a good wagon road in summer, and saddle and snow shoes in winter."

24. It is regrettable more is not known about Miss Robins. Women were not commonly involved in mining and smelting at this time. See footnote 22.

THE MAN AND HIS TIMES

*Photographic Essay*

$\mathcal{E}$ ben Olcott traveled to the San Juans as a young, but experienced, mining man. Belying the often pictured image of the "rape and run" mining industry, he had a reverence for nature in the world about him. This reflected a growing trend among Victorians in the nineteenth century, even among those out west developing their promised land.

His era can be recaptured in two ways. One through photographs of this the first generation of Americans that fell in love with the camera and preserved the images and age. The other is to hike into the heart of the San Juans, upon King Solomon Mountain, to see what Olcott saw and savor in your own way this wondrous land. There is not much left of the mine, but the scenery has not changed dramatically in the nearly century and a quarter since Eben lived and worked there.

Even Silverton, and to a lesser degree Lake City, retain some of the flavor of the 1880s. Eben Olcott might even recognize some of his familiar haunts.

Eben Olcott.
*Courtesy: New-York Historical Society*

Lake City probably in the late 1870s.
*Courtesy: 1st Federal Savings*

Silverton 1885 c. The North Star mine perched high to the north, to the left of the photo.
*Courtesy: San Juan County Historical Society*

Olcott stayed at the Walker House in Silverton.
*Courtesy: San Juan County Historical Society*

The Crooke Smelting & Concentrating works was the San Juans' finest for a brief time.
*Courtesy: Frank Fossett, Colorado, 1879*

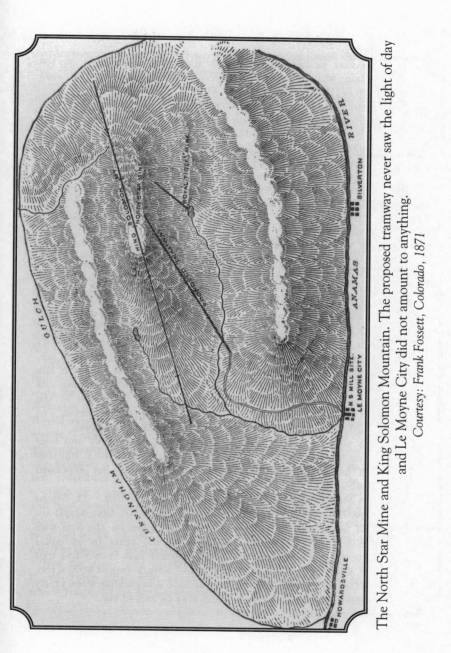

The North Star Mine and King Solomon Mountain. The proposed tramway never saw the light of day and Le Moyne City did not amount to anything.

*Courtesy: Frank Fossett, Colorado, 1871*

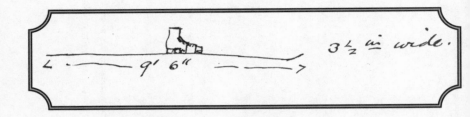

Two of Eben Olcott's drawings that he included in his letters
showing his "snowshoes" and how the men shoveled out the snow.
*Courtesy. Duane A. Smith*

North Star Mine, 1880c probably about the time Olcott first saw it.
*Courtesy: San Juan County Historical Society*

Isolated and high, mule trains had to supply the North Star.
*Courtesy: Flora Dowtain*

The North Star had expanded by the mid-1880s into one of San Juan County's more promising properties.
*Courtesy: San Juan County Historical Society*

The mining
crew proudly
posed in
the 1888
photograph
*Courtesy:*
*San Juan County*
*Historical Society*

The Silverton hiking club reached the mine about 1890 probably bringing more women than had ever been, or would be, at the North Star. *Courtesy: San Juan County Historical Society*

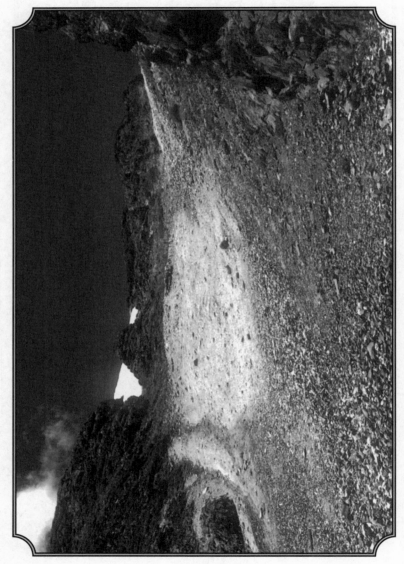

Once there were buildings, miners, mules, the Silverton hiking club, and Eben Olcott here. Now nature is reclaiming its own. *Courtesy: Duane A. Smith*

A few boards, some pieces of glass and the iron rods and cable that kept the buildings from flying off the mountain are all that is left of the North Star Mine in 2001. *Courtesy: Duane A. Smith*

❈ 63 ❈

The pack trail to the North Star came up from the lakes at the right of the photo. *Courtesy: John Ninnemann*

The trail to the North Star curved around the peak
in the center.
*Courtesy: John Ninnemann*

The views in and around the North Star are every bit as
breathtaking as Olcott described them.
*Courtesy: John Ninnemann*

A lone claim marker on King Solomon Mountain reminds visitors that mining once briefly reigned here. *Courtesy: John Ninnemann*

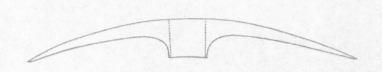

## CHAPTER 3

# *"You can't tell what surprises may be in store for you"*

*B*y crisp early September, with fall already setting in, Eben Olcott returned to Lake City. He wrote Pheme on the third:.

*[Sept. 3rd, 1880]*

*My dear Pheme*

    *I have a good deal to write but very little time left at disposal. I had a most perfect trip out here. Wednesday night I left Alamosa on top of the stage coach and despite my duster and lap robe I was quite cold and had to exercise all I possibly could to keep warm enough. The driver's hands became so cold that they were numb, that is quite a change from the heat of Chicago last Monday. The scenery all along the Rio Grande and also in Clear Creek is surpassingly beautiful. Though I have been over it a number of times it always appears more exquisite but it so varies as to defy description at least from me. I made do entire from on top of the coach. Last night I arrived here soon after eight and today have been busy with some preparations for my trip over the range. I shall not start until Monday morning and from that time on I expect to be very, very busy. I have quite a number of little business trips to make and will have enough at the North Star to keep me very busy.*

    *The weather here is now exquisite. I do wish it would keep so instead of going way down till it freezes the mercury as we may expect before very long. I have had quite a horse back ride today. I was greatly delighted at receiving your note together with two other letters so soon. They came tonight. I did certainly miss my pistol woefully. I certainly thought it was put into my valise and I hunted for it everywhere before I left Alamosa. You need not however have sent it prior to the boxes. Give lots of love to all. I will write again as soon as I can.*

The next letter came from Silverton, amid a San Juan autumn with all its splendor. Even with his busy schedule, he had time to appreciate nature's beauty—but then he always did.

*[Sept 9th '80]*

*I received yours of 2nd tonight. Although it is really but a short time since I heard it seems quite an age. I have been doing so many things preparatory to starting up that frozen up mine that the days seem interminable and at night I don't know exactly where I belong and have no nook that I can call my own. I am scouring the vicinity for a house till I can build one and for the most part I am in a hotel where they open their eyes at you wanting any better place to wash than their common sink off the office and think you are stuck up if you ask for a bed alone not to say a room.*

*I am a little sorry all my things are coming in one box. I will have to have them separated as my boxes have to be hauled by pack animals so they must be small. Never mind however.*

*I feel as though I wanted to have about twenty hands but more particularly twenty legs to carry me in the different directions I ought to be in. I shall have to take time enough away from here to go to Lake City in a day or two.*

*You must not expect much in the way of letters until I get settled. Write to me as often as you can.*

*The weather so far has been magnificent. Only a bit cool today. We had another fall of snow but it did not last long. It is so cold even down here that in some bedrooms you can see your breath freeze. I would not mind these settlements or Lake for the winter and hope to get somewhere besides the top of this mountain.*

*There are some most beautiful things about this country. The mountains today look superb with a light covering of snow. The majority of the trees here are mountain spruce but there are some aspens which are now yellow and red with the autumn change and contrast most picturesque with the dark evergreens.*

*Hope the Ponds[?] and ink will agree — my gutta-percha ink stand leaked in my trunk but did not do a great deal of harm — a few underwear, a flannel shirt and a handkerchief are somewhat stained.*

By September 14, matters began to settle down a bit. Eben wrote his mother a letter, most of which dealt with family matters. It does, however, provide some insight into his relationship with John Crooke.

*I was very glad to get your letter of the 6th. Tonight on my return from a hard days work up Picayune Gulch I feel as if bed were the only tolerable place for me and yet I must make up a lumber bill before I can retire. Your letter was most interesting. I am so glad you have had two such pleasant visits to the country this year. I wish you could spend the entire summer out of the hot city. I wish we could import some of your superfluous heat to these mountains for even now we could stand it without any discomfort. It is a most magnificent night. The moon shines almost as bright as day and there are more people in the street than if it were day light, which is often the case about mining towns.*

*Tonight I received a pleasant letter from Mr. Crooke which I think you would like to read; it is most characteristic; you will see he is peculiar and very pleasant when he wants to be. I am terribly hampered in my work by lack of pack animals. I don't know what I shall do if I don't succeed in finding more shortly. Every minute of my time it takes up and I have little time for letters.*

He concluded that letter with his opinion about women traveling alone. Apparently, this was brought on by one of his missing letters, which included something about three women who were "in the same sleeper with me. I find ladies traveling alone almost every time I come out or go anywhere. I do not however think it desirable."

Olcott must have gotten many letters regarding business matters that he did not save. A sampling of those that have been preserved provide, at least, an indication of the correspondence.

While in Silverton and at the North Star that September, Eben received letters and information about the Lake City end of the business, usually in response to something he had sent over or questions he had posed. A few were quite short, for example, a list of assays, dated September 14. The assays proved encouraging running from 18 1/2 to 51 ounces of silver.

John Crooke wrote a complimentary letter four days before, full of Victorian sentimentality: "I predict for you a great success. I predict because I have an unlimited faith in the mine." Crooke was also pleased with the syndicate. "When I reflect upon the syndicate affairs, a feeling of profound satisfaction warms my heart, and fills my soul with gladness."

Somebody must have criticized the mine's potential, or development work, or been lukewarm about it. That did not please Crooke in the least. "Those dogs gave only the tribute of a sneer to the best and loveliest property in San Juans. It was a case, a practical case, of 'casting pearls before swine'." What most excited John was one sample of ore which carried 563 1/2 ounces of silver and one ounce of gold. Then he went on to say, giving an epitaph to many a western mine, "It is a queer mine and you can't tell what surprises may be in store for you." He closed in a rather unusual manner and with an odd sentiment: "If you keep on being a good boy your 'uncle' will with permission granted, drop you a line occasionally." [1]

Regrettably, more of Crooke's letters did not survive. They would provide further insights into the Olcott/Crooke relationship, company observations, and the personalities of the Crookes. Some hints come from Olcott's letters; John's letter only teases the reader.

Olcott needed to keep in steady contact with the smelter staff on a variety of topics, mainly assay reports. In a September 28 letter, regarding samples, the assayer wrote about a smelter incident. Though it had nothing to do with the mine, it provides another example of the perils of western mining.

*About a week ago two of our furnace men got in a fight & after it had been settled one of the men (the one that was whipped) sneaked up behind the other one & hit him on the back of the head a deadly blow with a stick of hard wood. Larry the one who is hurt is still alive but hardly think he will recover. The other Bill Henderson is*

> *in the calaboose. There has been some talk of lynching*
> *but I guess nothing will come of it.*

The next day, Crooke wrote a letter but did not mention the incident.

The *Silver World*, September 25, carried a short article about the "affray" which took place in Bergin's saloon at Crookesville. [2] The victim, Lynch, had his skull fractured by the blow and, as, the letter mentioned, he was not expected to live. "Whisky was at the bottom of the quarrel," the report concluded. The next issues made no further mention of the incident or what happened to the two participants. Apparently, though, no lynching took place. More important matters like politics and the dedication of the new school grabbed the headlines.

What Crooke wanted to do about working the mine that winter entailed some discussion. Having just returned from Rose's Cabin [3] and "the mail [departure time] is about to close" he proved brief, however: "In my opinion the mine should be worked this winter and that you had better arrange matters to that end. The assays you will find below. There is rich ore in that mine and no mistake."

On October 1, Olcott received a letter from Lewis Crooke, the company's secretary. Eben had apparently written him for instructions regarding accounts for labor, sinking, stoping, supplies, and the like. Lewis wanted all returns sent to New York, seeing no "necessity for any of the company's accts. being kept at Lake City." He went into some detail about accounting matters and procedures, so that the "actual cost of development and of the ore, which you may send to Lake City, can be ascertained." Crooke then concluded "please render me a statement of your expenses and salary."

Although distance remained a problem, Olcott was blessed to be connected with the best smelter in the San Juans. Frank

Fossett's 1879 edition of his nearly encyclopedic guide, *Colorado*, highly praised the Crookes and their smelting and mining efforts. The "first eastern capitalists that showed their appreciation of the region by putting their money into it" conducted a smelting business in New York City, according to Fossett. Their investments in the Little Annie and Golden Queen Mines and Mills had led to an investigation of the Lake City silver district. The Ute and Ulay Mines, west of town on the Engineer Pass road, had been their first purchases—and most developed properties— by the time Olcott arrived. Their smelting works, with the "finest water power in the State," were completed in 1878. "Up to this time Crooke & Co. had expended over $400,000 on their mines, works, and other property of this locality." Fossett had nothing but praise for their efforts, and, in truth, finances and smelters were just what the San Juan needed at this juncture of their mining development. [4]

Fossett also explained the relationship between the two companies that the Crookes had created. Olcott technically worked for the North Star Mining and Smelting Company which had McPherson LeMoyne as president, Lewis Crooke, as mentioned, was secretary, and John Crooke was one of the trustees. The company owned the mine, a mill site, and part of an unidentified town site. The Crooke Mining and Smelting Company (1879) had Lewis Crooke as a trustee and John Goodwin as president, who also served as vice president of the North Star. Lewis had overseen the construction of the smelter, and he and John were closely involved with its operation and hoped for its success. [5]

According to Fossett, the Crookes planned to build the North Star smelting works when the Denver & Rio Grande reached Silverton. They optimistically predicted the line would come within two miles of the mine. Money had been raised for the project; Olcott had heard a similar promise.

With perchance a misgiving or two, Olcott climbed back up King Solomon mountain and sent two revealing letters from there to Pheme.

> *North Star Mine*
> *King Solomon Mtn.*
> *Sept. 19th, 1880*
>
> *I have yours of the 10th from Caldwell to answer. I received it yesterday it takes a full week for letters to come to Silverton for it is more mountain locked than Lake City is. From a mining point of view Silverton is however much more richly endowed than Lake is. The mountains are all mineral bearing and certainly more interesting. In searches I have made for fluxes I have been far more successful than I could have hoped for and found most valuable specular iron [6] and flurspar [7] nearly side by side. The smelting co. who have been here for years do not know or at least do not use either and I would not let them know how valuable I think them for hundreds of dollars. [8]*
>
> *Do you understand the rights of U.S. citizens in locating unoccupied mineral land? They can when they discover a new vein or deposit locate 10 acres of it [on] a piece of ground 1500' x 300' and if they do $100 work of work on it annually record it and conform to certain other requirements of law they can own it and after $500 worth of work is done on it provided no adverse claim is entered they can get a U.S. patent for it. [9]*
>
> *If a claim is located and the annual assessment is not done on it can be relocated or jumped by another party.*
>
> *This provision of the laws stimulates the army of prospectus who are constantly scouring the mountains.*
>
> *I am unable to describe the one hundredth part of the glory if some of our sun sets from the lofty pinnacle. I was coming up the other evening and just as I got*

*where I could command a very extensive view the sun was approaching the horizon. It had been snowing a little and the heavy white clouds floated all around me throwing columnar masses down into the valley beneath which spread out at their bases more beautifully. The loftier clouds were constantly changing from one glorious hue to another when the sun appeared and lighted up the whole scene with a magnificent base color spreading through every cloud, casting its reflection on the dozens of sun sprinkled mountains, making as it were the whole air red. It was enchanting, wonderful, surpassing description. Last night as the sun sunk in glory the moon rose with its silvery light and all night was as day.*

*I am trying my very most to make a success of this enterprise and have met both with encouragement and reverse. I have not been here long enough to say positively and at any rate this year is no criteria as we started so very late. I am building a nice little bit of a house down in the valley as well as a stable. When I am done there I will have to ride into town 2+ miles, to all my meals which will be a nuisance but the best I can do at present. I expect to spend some time here.*

*Today the men were having a perfect rest which they are enjoying.* [10] *The day is unusually fine.*

*I have had such long interesting letters from Frank lately real shop news. He is a brick. I have just written him 13 pages and am rather disinclined to continue much longer. I have an excellent Capt. or Foreman one of Crookes men. On many accounts I would prefer Sam Skems but as the time I will require him is short it would not pay him to come out. Such a lot of bad characters have come to Silverton that the town's people have formed a vigilante committee. It was very necessary as they were waging things with a pretty high hand. [11]*

*My box has not arrived yet here. I hope for it soon, as yet I have no place where I could make use of the many things. I hope however soon to have my little house built where I can arrange matters more to my liking. I scarcely know how to get my things transported over to Lake when I get through here it is easier to move hundreds of miles in the east than over one range of mountains here.*

*North Star Mine*
*Sunday, Sept. 26th, '80*

*I don't think I have much to write. To finish up this Sunday that has just flown. This morning Cap and I spent in walking the mountain. We did not get in till an hour after dinner. Then I wrote and read till tea time by our flickering candle light. I do not wish to write a great deal. I have been hoping for my shades [12] which have not yet put in an appearance. My mail has been pretty light lately. Thus I do not know but what I expect too much, home letters have come in pretty good numbers for which thanks.*

*Last night it snowed and although it has been bright today the snow has scarcely melted. We think that quite a cold winters day at home and yet here we are not rid of Sept. I am vexatiously delayed in my preparations for winter by the great difficulty in getting lumber for my new mine house up here. Just think of building having to pack every article on the back of donkeys, we have to twist and turn and invent to make a doubly strong structure not of very light lumber. If real cold weather and deep snow set in early I certainly cannot make the preparations that are necessary for wintering the snow up here but if as the weather-wise say we are going to have a late winter we may succeed in having a good deal of work done during the winter months. I think even if I have to stay out here all winter, which is probable, that it will seem shorter than it otherwise would by being broken by my change to Lake City in November, then by my return here as early as practicable in the spring. I only want to see the concerns making money to be very well evaluated . You see no doubt that there is a new Supt. at the Chrysolite. [13] He is one of the S. of M. boys who graduated the year after I did. I know him quite well. [14] If he succeeds he will make a big name but the position is a rather uncomfortable one at present at least. Is Mrs. Dekay blue about their stock? She has good reason to be. [15]*

*This week has passed very rapidly and I have been tied right down — or rather up here with the exception of one day. My little house below is about completed and now I am waiting for the things that were shipped out of order to make it habitable.*

*I am very well, one of my men told me yesterday that I was as stout — which he meant strong — as any man I had. I am not sure but what that two months experience may be very good for me.*

Nature was about to provide the most serious reversal Olcott encountered. The winter of 1880-81 was one of those that pioneers remembered vividly, a little more fondly later, but not amidst mounting and drifting snow. Despite some encouraging signs, the transportation and isolation problems still hampered North Star operations. The D&RG platted a new town south in the Animas Valley, Durango, but the tracks had not arrived, and there was little hope they would reach Silverton in 1881. The only visible appearance of the D&RG

were track-laying crews and other employees who wintered in the infant settlement.

Durango, though young, had quickly emerged as the largest town in the San Juans having topped 2,000 population with the aid of the D&RG people. It also had quickly become the largest community on the Western Slope of Colorado, that land out there beyond the Continental Divide, or the "snowy range" as they called it.

The Crookes also remained evasive about the Boulder Gulch Mill. Probably because they had spent large sums of money, and the returns had not measured near to expectations, the brothers and their partners were becoming concerned and not digging so deeply into their pockets.

*The WalkerHouse* [16]
*A Home for Families.   -1880.-   The Travelers Resort.*
*— W. S. WALKER, Proprietor —*

*Silverton, Colo. October 11th, 1880*
*At these times when it is impossible for me to write frequently I almost think my daily letters are apt to make you wonder where they are.*

*I have been fighting snow for over three days. Friday I had a government employee up at the mine to take down statistics of our production. He was an S of M man and most agreeable and we came down together. I little dreamt that I could not go back when I wished to but the storm was so tremendous that I only got half way on Saturday and Sunday (yesterday). I spent eight hours of tremendously hard work trying to get two miles and failed and had to come back to my place of starting.*

*Today we did all we could do to break the track from the halfway cabin to town . It is still snowing hard and has been for 76 hours at the mine — not less than*

*3-1/2 feet on a level has fallen and here there is over 2 feet. Tomorrow I shall go back with Calvin and hope to get up on Wednesday.*

*I have enjoyed this storm personally but it is very disastrous to my work. I will write when I have time again. Everything in my box seems in good condition.*

*Silverton, Colo.. October 12th, 1880*
*[Again writing from the Walker House]*

*Still snowbound — it has only been storming hard for 98 hours 20 minutes and still she has not given out. The town is full of storm-staged strangers about half of whom (judging from appearances) are half seas over.* [17] *We had a most amusing one as our table topic at supper. He was just jolly and shameless as it was we had to laugh at some things another might tremble. Another rough tumble miner with torn coat when he saw one of the guests looking at his hands he said 'You need not look at my hands. I have not washed my hands since the 22nd last April. The alkali water is bad on them.' We sit, I want you to understand, at the first table and those are his sort of men that occasionally drift in to make up our mining life. Now I am in the office, a room about 20' x 20' in which are about 27 men, nearly all of whom are smoking, a big air type stove is turning out its heat upon their wet coats of the occupants. A big tin box filled with snow is on top of the stove melting to make [the] water we have as ditches and water carriers wagons are frozen up. The aroma in this room is delicious as you can imagine.*

*Although my house is complete I have had nothing to put in it yet and the last few days the roads have been so bad that I could not haul a stove and wood to it. I do not object to these things either. Once I did now*

*I have got used to them not so however but what I would enjoy refinements more.*

*Saturday night last as I believe I wrote you I was belayed and had to stay at the halfway cabin — I had no blankets so had to sleep on the hardest of bunks wrapped simply in my overcoat. Towards morning it was frightful cold and I wished and wished for morning to come. Still I took no cold at all!*

*Sunday I had an awful day. It was one continuous fight with the snow for 8 hours and when I got back to the cabin I was glad enough to eat the fried beef, tea and bread which tasted delicious to me. My appetite is splendid and I feel very, very well and strong. I think I have improved a good deal in this respect since I have been out this time. I think tomorrow I shall make an order although to go to the mine tomorrow, probably in snow shoes.* [18]

*[Drawing of skis]*

*I don't know what sort of a figure I will cut in them as it will be my 1st attempt.*

*I think probably we will after this storm have a [period of ?] good weather during which I hope to accomplish a good deal of work at the mine.*

*The mail has not got in for 2 days; hence forward you need not be surprised at irregularities in our mail as there is a very difficult range to cross which is apt to get snowed up. When I get to Lake I will be writing a telegraphic communication at any rate and probably the mail will be regular. I now have a room here to myself and how deliciously I do sleep. If I can only make anything like a success out of this North Star Mine it is all I ask it will be worth bearing the few inconveniences there are.*

Eben did finally make his way up to the North Star mine, and he wrote Pheme from there. The weather remained a paramount issue at his mine.

*North Star Mine*
*Oct 17th '80*

*Sunday evening finds me pretty tired for after eight days of punishingly cold and stormy weather we found ourselves blessed with a day when outdoor work was possible and I have spent it making necessary preparations for the men's comfort and safety. I was down as far as timber line this morning and this afternoon have been shoveling snow out of the house and trying to stop up chink holes all over. Last night the wind abated and the snow which has been drifting in blinding sheets was nearly at repose and we have begun to take heart again. Yesterday was the coldest day I ever felt and it was almost impossible to go outside. I was really afraid some of us would blow away so today I had ropes stretched along the trails and now I feel a good deal easier.*

*The snow seemed to blow right through the boards of the house or at least take advantage of the minutest cracks. My conscience was never easier over Sunday work and I am thankful we have accomplished what we have. I came up on Friday as I wrote you and had a very pleasant trip but soon the wind commenced again. It was and indeed is a magnificent sight to look out upon this white sea of mountains. The full moon and Jupiter have just risen to add their brilliance to the picture and you could almost see to read out of doors. Last night I woke up way before day and thought the sun must be up as our room was so light. I can give you no idea of the mountains of snow that we look down upon, from a few feet in front of my door for over 500 feet is an almost unbroken slope of it and 200 feet down you see a little corner of a house that is all but buried.*

*This is quite a respectable building on one of our levels which will accommodate several hundred tons of ore. Now you might forget there was such a thing as wind while 24 hours ago the gusts swept down upon us with such fury as to terrify one even though he knew he was in a house anchored and chained to the solid rocks. Yesterday two of the men who had only been outside half an hour came in with tears rolling down their faces with the intense cold. I had made them wrap up so that none of them have been frost bitten yet and I think with proper precautions they can all avoid it altogether. None of us ever go out in the snow without colored goggles to protect our eyes for when the sun is shining it is so dazzling as to be apt to blind one. Today it has been a pleasure to be out and I trust it will so continue as long as I am here. The lake that is in view from the house is a beautiful green body of ice which glances the sun's rays back defiantly and two others about as far away are completed covered with snow.*

*Last week I spent an evening with my Georgetown friend Theo Schwartz. ME [19] He was down here examining some mines and happened to be in town the same night. He had just come from a visit to the Little Annie Mine and we enjoyed comparing notes.*

*I expect to go to town in the morning and I hope to find some mail there, for ever so many days we have not had any and a little news from the outside world will be very welcome. It was rumored on Thursday when I was down that a telegram had been received in Lake City announcing a Republican victory in Indiana which I hope will prove true.* [20]

*I am very dubious whether I can take even the contents of my big box over to Lake. I shall make a desperate attempt to do so for I want to do lots of solid reading there this winter. The snow has detained the mails so that although we are but 33 miles off I have not had any communications from there for more than a week and I know business letters were sent from there, so you see if mails cannot get through what a poor chance there is for heavy boxes.*

*I am not sure I wrote you that the blankets had been received. They will answer my purpose very well. Thank Father for them, they are cheaper than I could have bought them for here decidedly.*

*It was most amusing to see the woebegone countenances of the dozens of men who were storm-stayed lately at Silverton and to hear their different ideas about the weather. The least ray of sun light brightened up everybodies' countenance and gave us a little hope that it would shine again.*

*On the whole I do not know that I object to my experience here. This fall it is at any rate most wonderful and seems to agree with me very well. I believe I shall make tomorrow my last trip to the valley until I go for good and will henceforth do my errands by proxy.*

*This living in two or three places is most annoying. I never have the things I want in the right place.*

The weather finally improved, and so did Olcott's spirits. At last, he was able to accomplish his move back to more comfortable, less isolated, Lake City, with its urban attractions. After one more letter from the mine, he departed.

*[October 22, 1880]*

*As mails are such uncertain things here as I have
an opportunity of sending a note over the range towards
Lake I will do so in case my letters from Silverton are
still blockaded.*

*The past week has been most beautiful and I have
enjoyed it intensely. I should like you to see the frost
effects — some mornings down in the Animas Valley
every tree, shrub and twig is covered with brilliant frost
with some leaves and crystals of it an inch long. The
snow also on the ground is ornamented with a similar
frosting in leaves not unlike those of copper I brought for
Ore Knob.*

*Twice I have been under the mountains just as the
sun rose above them and began to slant down their sides
through these frost ornamented trees and it made them
glisten most beautifully. It also has the effect of loosening
some of their crystals, i.e. which are so light that they
float in the air making a perfect cloud of spangles which
shine, reflect and decompose the light till you have every
color of the rainbow.*

*The sun rise and set effects are wonderfully varied
and beautiful and we gaze on them always, the Capt.
being quite as great an admirer of nature as I am.*

*As I wrote you lately please send all letters to P. O.
box 445, Lake City hereafter. I will probably be there by
the time they arrive. I am anxiously looking for a large
mail today but may be disappointed.*

The letterhead on Olcott's new stationery was the first time
that he has used it, and it showed he still maintained his New

York office. The letter is addressed to "My beloved sister," presumably Pheme once more.

*E. E. Olcott, E.M..*            North Star
  *Mining Engineer*       Mining & Smelting Co.
                              SILVERTON, COLO..

Eben E. Olcott, E.M.
  *Mining & Metallurgical Engineer*
*Agent, North Star Mining and Smelting Co.*
*Silverton, San Juan Co., Colorado*

*111 West 13th Street   New York*

*Nov. 4th, 1880*
*I have been delighted at receiving many letters from you lately. You are very good and I am sorry you have been anxious over me. There is no cause for it for if other people can endure cold why should not I be able to? I am alternately freezing one half or other of me tonight for the stove has no effect upon that side turned away from it. I don't know where the mercury is now, much less where it will go before morning but it is awfully cold. You have no idea of it. I am in the little mite of a house I put up and only one of my stoves is up so I am scarcely prepared for the severity of the weather. Fortunately I will sleep comfortably and to my most comfortable little couch from Hall & Stevens I mean to retire soon. We are almost through here. Capt. and I will probably start next Monday (8th) for Lake City, Of course, a foot for saddle animals are almost out of the question now.*

*I should like to introduce sleigh dogs here. I see no reason why they would not be of great benefit except for the snow slides over which if one must pass Norwegian snow shoes are undoubtedly the safest things to use. I can manage them pretty well but of course need practice to become an expert. Today I was very glad to get my black prinsers though I wanted them sent to Lake City. It will be a nuisance to carry them over and I won't trust them to the mails again at this season. You have no idea how careful we are in cutting down the few articles we must carry in these long tramps over the snow in walking 15 to 18 miles every ounce counts. We expect to accomplish 18 miles on the journey to Lake on Monday, stop at Roses Cabin overnight and make the remaining 14 or 15 miles on Tuesday. It is so much more comfortable at Lake and we have telegraphic lines and very much more of a feeling of civilization pervades the place. Last winter the mail only missed one day.*

*I find the valley here on the banks of the Animas colder than the tops of the mountains when there is no wind there the sun gets down there so much later. If it storms Monday afternoon we will not be able to cross Engineer Mt. and may have to wait some time at this season for a clear day. We have not yet heard any authentic reports from the election. Here we did nobly for the Republican cause. [21]*

*Together with the other packages received today tell Jake I got the two rubber stamps and the ink; Cap. also received his garments from [Mg. ?] for which we are very glad for I was afraid they would not be here before we had to leave. Thank Jake for the stamps. They are just right, as you see I also received my paper from Laurier. It is all right and was well done up for our rough mail carriers to pound. I am quite sure I wrote you that Ink and Ponds were received in good order, but as you excited yourself to a panic [?] there must be no doubt hanging about the subject.* [22]

*Cap. is playing fireman so well that really we are thawed out and I guess will keep so till morning.*

*. . . I received the wristlets* [23] *a day or 2 ago. Don't know but I shall have preferred the silk ones you refer to. I do so abominate myself in a flannel shirt and the lack of linen cuffs is a constant eye sore to me that I like to cover up the unsightliness a little. The house items I was glad to get. I think everything is moderate. You might have thought that it were possible to be lost in snow drafts if you had seen the track we had to shovel out. In places the snow is 10 feet deep and I had to work men on two separate benches in shoveling it out.*

*[Drawing of men shoveling]*

*Your letters are all worth thousands to me and I hope you will continue to write as often as you can..*

*When I get to Lake I expect to be very busy but will let you have a word or two now and then.*

Whether or not Olcott was too busy to write, there is, unfortunately, a two month gap in his letters. Meantime, winter set in, and the high San Juans were again trapped by cold, snow, and wind. Even Lake City, on the outer edge, could not avoid that fate.

1. Crooke to Olcott, Sept. 10, 1880, New-York Historical Society. All the September and early October material is found in this archive.

2. In this small hamlet that grew up around the smelter, Olcott roomed with the Crookes during the winter of 1880-81.

3. Rose's Cabin was on the Engineer Pass road. It was a combination stagecoach stop, saloon, office, stable, and, for a brief period of time, had a post office. A "hotel" sat across the road. A little mining was done nearby.

4. Frank Fossett, Colorado (New York: C. G. Crawford, 1879), 516-19.

5. Fossett, *Colorado*, 519 & 526. *Silver World* (Lake City), June 3, August 19, 1876, & June 12, 1880.

6. A variety of hematite, containing 70% iron. It has a black, lustrous, metallic gleam. A flux helps promote the smelting process, for example, absorbing mineral impurities in the ore.

7. A natural calcium fluoride, florspar or flourspar, is used in gold, silver, copper and lead smelting.

8. Olcott must be referring to the Greene Smelter, which was Silverton's first in 1874. On the north end of the town, it had a checkered career before most of the equipment was moved to Durango to become part of the San Juan & New York smelter in 1881.

9. Olcott is alluding to the 1872 mining law, still in effect in 2002. He clearly described the two types of lode claims, the unpatented one which requires assessment work and the patented claim that is owned outright. Based on earlier mining district laws, the federal law unabashedly encouraged development of mineral resources. The law stated claims could be 1500' x 600', not Olcott's 300'.

10. Olcott is writing on Sunday, September 19. Miners typically worked a 5 1/2 or 6 day week. Barring some type of emergency, Sunday was the day off. Standard wages were $3 or $3.50 per day.

11. Allan Nossaman makes no mention of vigilantes in either Volume II or III of his *Many More Mountains*. One incident of a shooting occurred on September 16, which may have caused the idea which Olcott mentioned. See Nossaman, *Many More Mountains*, vol. 2, 304.

12. Apparently, Olcott had ordered some type of window shade and was not using the modern slang for sunglasses.

13. Olcott is apparently referring to Leadville's Chrysolite mine, one of Colorado's premier mines until June-July, 1880, when its stock price broke disastrously. Everybody blamed everyone else in this fiasco about insider information, stock trading, poor mining practices, and a gutted mine. Coming on the heels of the Little Pittsburg failure, Leadville's glamorous mining days ended.

14. He may be alluding to Charles Rolker who became general manager of the Chrysolite on September 1.

15. She had reason to be worried, indeed. Chrysolite stock collapsed from $23 in late April to $3.75 by early October, amid similar accusations to the Little Pittsburg mess. Both mines were finished as major producers. DeKay, perhaps a husband or relative, resigned as secretary of the Chrysolite Silver Mining Company in the midst of the collapse.

16. On the corner of Snowden and 13th St., the Walker House was Silverton's finest. Built in 1875, it was also noted for its good food. The owner's (William Walker) congeniality "gave the place a glowing reputation." Nossaman, *Many More Mountains*, vol. 3, 319.

17. Nineteenth century usage for drunk.

18. He is referring to skis, as the sketch included in the letter clearly shows. Calling them snowshoes was typical.

19. Theodore Schwarz (Nossaman, *Many More Mountains*, vol. 3, 20 says the correct spelling was Schwarz) would return to the San Juans to become superintendent of a number of mines during later Red Mountain excitement. He was active in several Colorado mining districts.

20. Republican Olcott must have been happy when the news was confirmed that James Garfield had been elected president. Coloradans elected Republicans James Pitkin governor and James Belford to the House. Pitkin owned mines in the San Juans.

21. Colorado's Western Slope counties, of which San Juan and Hinsdale were the most populous, gave Garfield a narrow thirty vote victory, Pitkin nearly a 400 vote margin, and Bedford lost by 200 votes. The eastern slope countries, strongly Republican, carried the day decisively. Eighty-six percent of the vote came from there.

22. Most likely Olcott is referring to his new stationery. The North Star Mining & Smelting Co. Silverton, Col. obviously was stamped on the letterhead.

23. Material worn around the wrist for warmth, like a cloth bracelet.

CHAPTER 4

*"The richest treasure*
*vaults of San Juan"*

$\mathcal{I}$n its January 1 edition, The Lake City *Silver World* greeted 1881 with the usual enthusiasm that a mining town editor mustered for the holiday and the new year. Referring to its mining region, the paper crowed with the typical boomer bombast, "these camps and districts comprise the richest treasure vaults of San Juan, the richest in the world." Nor did the paper stop there, the future glowed rosy: "Such is the present. The future is rich in promise, is fraught with great and powerful results."

Eben Olcott went home sometime in late November or early December for the Christmas holidays, which explains the initial lack of letters. On his return, he left some materials in New York that he later would need. On January twenty-third, he asked his sister, "I recd. it while at home & it was probably dated near end of Nov. from Del Norte. ...contains some mine matters North star & Shenandoah & Yellow Jacket also a sort of acct. I rather think I left it in the envelope & put it in the tray of the big truck in closet."

When he returned to Lake City, Olcott perhaps did not write very often because of abominable weather! The November 28, 1880 *Silver World* ran a column on weather news across the state during a cold, snowy late fall which would become a colder, snowier winter. Saguache registered sixteen below, Pitkin thirty-five below, and Central City twenty below zero. Middle Park had six feet of snow, and Colorado Springs had dropped from sixty degrees in early afternoon to snow by evening.

Lake City church goers were greeted with thirty-five below on Sunday, January 9, 1881. Crookesville [1] topped even that with forty-two below. The editor laconically noted that most of the local wells had frozen solid.

For several months there had been a scarcity of news about the Crookes and their operations from the newspapers. This ended in January when the *Silver World* carried two articles the second involving Olcott. As superintendent of the Crooke M&S Company, he had "taken charge of work and operations here." The smelter had been closed for "the purpose of making extensive additions" and "probably" would not be started up before May or June. The article then went on to say that John J. Crooke "retires from active charge of the operations of the company."

Someone misinterpreted what was said or happening, which caused a letter to the paper on February 4 from Olcott. He did say changes were being made at the smelter "with the view of increased capacity and economical reduction," but Crooke had not retired. The changes mentioned in the previous article apparently were correct, for they received no comment, nor did he refute the statement that the plant would be closed for several months. [2]

In a January twenty-third letter to Euphemia, [3] Olcott took up the subject of weather. Getting up in the morning proved invigorating, to say the least. Yet Colorado's dryness tempered the cold, allowing outdoor work.

> *We have had a very cold week the mercury has been around 20 below every morning. It is rather strange we do not feel it more, but I am dressed exactly as I was at home excepting that I never wear an overcoat except in driving & yet am not cold. It takes a great exertion to get up in the morning & dress which I always do before seven. It is a rare thing to find more than a bucket of water left unfrozen just enough to take a dash with. Nail brush,*

*tooth brush & occasionally hair brush are frozen stiff also
my towel if I have left it at all wet the night before.*

    *By the time I have been out around the works I feel
plenty warm enough & then I return for breakfast and
the rest of the day is very pleasant. The mercury goes
down in the neighborhood of 0 just after 4 P.M. yet we
work about in the evenings as unconcerned as possible.
Mrs. & Miss Crooke do not seem to mind the cold at all
and we are very comfortable in the house and get quite
used to ice.*

To make matters worse, when getting dressed, "in the morning
our shoes are also frozen stiff so that they are very hard to put on."

    Cooking in the mountains could be an adventure as many a
man and woman, and Olcott found out. Eben went on to regale his
sister with an experience Mrs. Crooke had with a turkey.

    *We have had a good deal of sport over a turkey at
table & I told Mrs. C. I was going to write about it. [It
was a large bird] & we all exclaimed at it the first day
it was on the table, but when Mrs. C. began to carve it
it turned out underdone & had to be sent away. That
P.M. it was further cooked & set away in the closet in
the kitchen to be served cold. So next day it was on the
table but frozen stiff so that it was with great difficulty
it could be carved & we ate very little of it. From this
beginning it has gone along till up to its final disap-
pearance it was on the table six times & Mrs. C. thinks
it would be a good plan of an economical boarding
house keeper. The bread has to be thawed out every morning in this weather.*

For all the problems with the turkey, Eben thoroughly
enjoyed staying with the Crookes. "I never staid at a house
(excepting here) where everything always tasted good as at the

Crookes table. They are great cake eaters which does not particularly take my fancy but the table is very plentifully provided with everything."

Working hard and long, he found few moments for anything else. "I find it difficult to find time to myself here. I never retire to my room except to sleep & my corner in the office is sometimes occupied & there are always so many little interruptions of people to talk to. Perhaps when we get rich I will put [in]a private office."

Olcott always enjoyed news from his family, commenting on it as he did in this letter: "Today I received your letter of last Tuesday and was as always very glad to hear from home. Time flies very rapidly but the days between letters from home seem like weeks." It was not always the fault of the folks back east. He told Pheme on February 6,

> We have had a rather provoking week in regard to mail. In the first place there was a wash out on the A.T, & S.F. railroad in Kansas and for a day or two there was no mail at all, & then there have been two coach robberies [4] in both of which all the mails were taken. Today I certainly thought I would receive a large mail but there was no mail at all from the east & we know not the reason.

To make matters worse, Eben was sorely disappointed in a church service he attended.

> I did not go down to the Episcopalian church this morning. We were all down last Sunday and were very much disappointed in the minister, [5] he preached a sermon that was very much out of place, then it was com-

*munion today & I never have been to an Episcopal*
*church for communion.*

*We expect to go to the Presbyterian tonight & there*
*the service is so very pleasant that it is delightful to go.*
*I wish they had two services. Really the sermon we had*
*last Sunday was a splendid discourse. I never heard a*
*man who used quotations from the Bible so aptly and*
*showed greater acquaintance.*

Despite his busy schedule, Olcott did find time for relaxation. He went on to explain to Pheme some of his winter
time activities.

*On Mrs. Crookes wedding day, I took all the fam-*
*ily up Henson Creek as far as the mine* [6] *& had a splen-*
*did sleigh ride. I wrote about it in a hurried note to*
*Anna but imagine the thieves got it. It is rather provok-*
*ing. Two evenings we went out coasting on quite a hill*
*& we can ride at least half a mile.*

*We made a long bob sleigh that holds six & one*
*night we invited two young ladies. It was quite a spree*
*but it came near ending disastrously. The sleigh ran*
*over a rock & broke a little so we could not manage it &*
*it ran to one side hit a log & threw five of us right into*
*the air several feet turning us complete somersaults*
*ladies & all. We all landed on our heads full & fair.*
*[no one was injured] We blamed the accident to a defect*
*in making the bobs & now have them remedied & I sup-*
*pose we will try it again.*

After happily discussing his winter sport, Eben returned to
the Crookes and the situation with the company. "Our [Negroes]
got fearfully impudent one day & raised cain & were discharged
forthwith; it was rather an expensive experiment. It is ridiculous

the way the C's waste money. I actually cannot write another word. Good bye."

That was the first sharp criticism of the Crookes in any of Olcott's remaining letters. Without question, they had spent money, some very wisely, and, at least in Eben's opinion, some quite wastefully. He was not, however, the only one to hold this opinion, and the financial outflow was not being matched by income.

He wrote a letter to his sister telling much more about his social life and times and continuing the stories of his winter adventures.

*[Feb 13th 1881]*

*The day is almost gone and I have not written to you at all though I had the best of intentions this morning to write you a long letter.*

*The week has been marked by the most exquisite-weather 17° to 20° below every morning but after nine or ten o'clock it has been much milder.*

*Last night we were coasting again everything was in prime condition and the full moon made it nearly as bright as day. Miss Nannie had her first experience on our bob and certainly shows much more pluck than Wm. I think the hill here is a good deal steeper than that at Clinton [7] and we fairly fly it makes it very exciting. I see in the last* Harpers Weekly *coasting is depicted as one of the common dangers that we are exposed to.*

*We have arranged to have Miss Cook up here for supper tomorrow and Jake and Wood will be invited in and then we will have a sleigh ride with four horses. It is to be a surprise to Wm. elaborate preparations have been made in the way of cake and pie. Miss Cooke does*

*make the best pastry I ever saw, puff pastes that are every
bit as good as Walls, and mince pies beyond comparison.*

*We were all down at church again and were very
much disappointed that Debble [8] did not preach, he
made a few remarks however. The church was packed
at least every seat was taken. It is a great satisfaction
to me that we have this exceedingly rare opportunity in
a mining camp.*

*The tableaux on Friday night were the source of a
good deal of amusement and passed off very successfully.
There was a large attendance and St. James S. S. [9] must
have done well.*

*Our mails have been very light again this week
though on Friday I received nine letters.*

*It is very late for me. Good night or will not get my
eight hours sleep. Love to all.*

A mining engineer always is looking for better equipment,
even candles. Olcott received a letter from Procter & Gamble on
February 19 regarding just that and promoting their product. It
gives an interesting insight into the times and mining.

*In making a test of our candles with those of any
other manufacturer we always select the best made, and
in a recent test made in Chicago, when the other candle
had but one inch left, there was still one and five eights
inches remaining. This of course is not much on a sin-
gle candle, but on a box of candles containing 240 candles it will amount to 120 inches, allowing one half inch
for convenience in figuring. ... so you can easily save
from 40 to 50 cts per box on the candles and this will
amount to considerable when used in mines which use*

*only a moderate quantity. Our candles give a white light and do not smoke.*

Whether or not, he chose Procter & Gamble is not known. Regrettably, other similar letters were not saved in the remaining Olcott collections.

At this point there is another gap, three months, in the Olcott correspondence. The long, harsh winter prevented work at the North Star, where it would be summer before operations started again. Olcott may have gone home, or more likely, continued to work at the Crookes' smelter. He also began again to be more actively involved in his consulting career.

One reason for suspecting he stayed in Lake City was the April 2 issue of the *Silver World* which had a comment from Olcott that a rich body of ore had recently been struck in the fifth level of the Ule mine. It ran 650 ounces of silver to the ton and the vein showed "probable continuation of that rich ore."

Work for the Crookes may have become routine, and maybe in short supply, which would have pushed Eben toward outside engagements. It would be several months before full-time work resumed on the North Star and the mountain mines. The smelter was still being overhauled, and the promising Ule, "the best developed and largest producing mine in the San Juan," according to local pride, had encountered water problems and would not be again worked full time until a sixth level opened. Miners would not start driving that level until the middle of May weather permitting. [10]

In his next letter Olcott highlighted some of the problems facing himself and Lake City. The mail robberies were getting on peoples' nerves.

*[May 22, 1881]*
*We have no mail again today. This week has passed quickly and uneventfully expecting the arrival of J.J. on Wednesday; he has started us on a little more*

*work than we have been doing. He is a man of many eccentricities as while some are not disagreeable and afford us much amusement others are decidedly objectionable not so much to me as to Mrs. Crooke.*

*Old bachelors usually have their crotchets. The Crookes you knew are just starting a factory for tin foil and capsules in Chicago. Mr. and Mrs. Henry Crooke arrived there the day before J. J. left bringing with them two work girls (lathe spinners) they traveled them 1st class and took them right to the Palmer House. It must have been like stepping into fairy land for girls getting $7 to $10 per week. It is only a sample of the way the C.'s treat their work people.*

*We had a serious mail robbery [11] on the road this week we lost considerable mail and I a package of some value which I needed very much. One of the passengers on the coach was wounded. Two of the three high way men have been arrested. If the state or U.S. government does not take some strenuous effort to stop such things it will become a common occurrence. We are getting very indignant and the first man that is brought into Lake and known to be connected with this high handed robbery will have Judge Lynch after him inside of 12 hours.*

*The weather here is perfectly charming now; often ice at night but warm in the day time. Have I written about my horticultural efforts here? We bought some plants and I made window boxes, these have done splendidly also, we have some hot beds with seeds planted in them these have not thrived as well but still they are*

*pleasurable. The bay-window in the dining room looks
very pretty indeed.*

*What do you think of things political? How does
Garfield seem to you? I am glad Conkling has been
beaten and hope he will not be reelected to the Senate.* [12]

*I very seldom go down to town now, except Sunday
evenings for church. Mr. Debble continues to interest us
very much.*

Meanwhile, the stage robberies were finally solved after
some more hold-ups and more passengers being wounded. Two
men were captured and taken to Del Norte. That same evening
a vigilante group took them from the jail and hung them. Olcott
had been right about what would happen to them once they had
been captured. [13]

Eben wrote the next letter from somewhere on the Union
Pacific Railroad line. He was off on a consulting trip to Nevada.

*[June 7, 1881]*

*I am enjoying the most exquisite trip. Perfect
weather most wonderful good fortune in coming across
nice friends every where – first starting out from Lake
the day was perfect the stage ride exhilarating and I tri-
umphed by driving the 4 horses a long distance. I have
ridden over the stage road over 1500 miles (14 trips 116
miles each) and I never have seen a passenger drive
before. I have had a good deal of experience since I have
been at Lake in driving which I have enjoyed extremely.
For you know prior to this I have done much more horse
backing. I had tremendous work to get ready to come a
way two nights. I was up till 2 a.m. so I left tired but
have rested en route.*

*I had hoped for your letter of a week ago Sunday
before I left but was disappointed as I only expect to be at
Galena (13 miles from Battle Mt.) about 4 days I did*

*not think it best to have my letters forwarded so there will
be quite an accumulation when I return. You have my
telegraph address in case you need me. I had no idea of
going when Porter telegraphed me but J. J. said he
advised me to go & they could get along without me so
off I started. I am going to examine the Aetna property
formerly White & Shiloh mine.*

*In striking the R.R. at Alamosa I met a U.S.
friend, at Pueblo another, in Denver I met my best
Colorado formed acquaintance T. E. Schwartz M. E. of
Georgetown with him I spent a pleasant half hour on
train today fortune has smiled on me. A Mr. Olmstead
and family [of] very nice people of Georgetown are here
and they introduced me to Baron Steck [14] of whom I have
heard much. He is largely interested in mines was out
here 20 years ago before most of the towns had sprung
into existence. A great friend of Arnolds (Secty. Penn.
Lead Co.) also. He and I now have the same section in
sleeper. Mrs. Olmstead's 2 boys are about 5 and 7 years
are models in every way bright interesting fellows and
absolutely obedient in fact perfectly trained and educated by everything being explained to them. The trip via
Colorado Cent. R. R. right up past all the Rockies,
Longs Peak was glorious. The particular attraction this
year about the country is its exquisite verdure and in
fact gaiety with wild flowers. The plains at the foot of
Sierra Blanca were just carpeted with flowers while its
snowy sides made a fine back ground. It was all so beautiful. You probably recollect I reported once before.*

*This takes me over the most picturesque portion of
the Continental route. Another joy is that I telegraphed
Frank to know if he could meet me at Ogden in answer
to which he said he could probably come to Echo[15] which*

*will give us nearly four hours together. We will talk like school girls.*

*Today we have seen a good many Antelope they are so graceful loping over the plains. I liked that article on Denver you sent me the growth of the city has been surprising and the Union Depot is beautiful.*

*The travel is wonderful, Colorado Springs, Manitou, Denver, all crowded. Now we have two sleepers crowded and 3 regular cars the same. Some day you girls must see these places.*

*The Snow Sheds along this road are very annoying to be rushed into a dark covered passage way 3 to 4 times an hour and kept in each 4 to 10 minutes is anything but conducive to good temper when you are writing or reading.*

Olcott generally enjoyed traveling, especially by train, and this trip, despite the snow sheds, was a joy. So were the fresh fruits he encountered.

*[Galena, Nevada]*
*[June 10th, 1881]*
*I have had quite a rush over the country today behind a pair of colts one of which has only been in harness a week this with the tramping through two mines and a good deal of surface walking has made me feel exhausted but as I do not get any home letters during this trip I must satisfy myself a little by writing to you.*

*You can not overestimate the delights of this trip. Starting from the mountains where fresh vegetables of any kind are rare and fresh fruit to come to within 12 hours of California to indulge in peaches, apricots, plums, cherries and strawberries and raspberries, ad lib, is certainly a treat. Then Strong the Supt. [16] here owns a large place at Los Angeles. (where I am going to take you some day probably in the grape season. 200 acres of vineyard) and receives oranges etc. by box so I indulge in the most delicious fruit ever so many times a day.*

*I visited here before at the most inclement season now it is beautiful. It rather frets me to see roses and other garden flowers in bloom and think of my Lake City plants having to be covered each night.*

*Strong met me at Battle Mountain [17] with his colts and we flew here. I will have to go down there on Monday to meet Porter – a millionaire E. N. Welch and Strong wants me to go alone, so I may break my neck but I guess not.*

*I expect to leave here on Thursday the 16th reach Lake on 20th. Ask Wm if he can send me one of those rolling blotters like he has in his room. I think they may facilitate the smoothing of the leaves of flowers in pressing I see so many beauties but never have time to press them decently.*

*I send you a genuine Ute arrow by this mail hope it will not get broken let me know if it does not arrive ok. The tip is probably poisoned. [18]*

On June 13, Eben sent his "sweet sister" a short letter from Galena: "Leave here to hurry to Lake City. I have had a very good and profitable trip and am glad to leave." He missed being in Lake City "very often, we are so comfortable it seems quite a long time since I left." Most of the letter dealt with family correspondence, but the first paragraph did tell a little about his work at Galena.

*What a luck you are I certainly was delightfully surprised this evening to have a 12 page letter from you put into my hands. I was very busy at the assay furnace conducting some very interesting experiments which quite surprise me. I was incredulous of a thing which Strong the Supt. insisted upon till I tried it myself.*

*Strong is a very fine man it is beautiful to see his devo-*
*tion to his wife and children or rather hear it for they are*
*not here but every evening he receives her letters and very*
*often there is some little flowers as memento in them. I*
*happened to see the heading of one 'My darling.'*

*I am fearfully tired tonight for I have been working*
*from 6 this A.M. till now long after 10 o'clock.*

Olcott returned to Lake City and, as he had predicted, was
overwhelmed by his mail. "I feel that I have so many of your letters to answer that I do not know where to begin." His letter, on
June 26, indicated that he did not plan to stay with the Crookes
but had not made up his mind what to do. Like its predecessor, it
mostly contained family news.

*We are having very hot weather something*
*unusual for Colorado. Last week in the stage the night*
*was very cold. I saw 3 of the planets and the most superb*
*meteor I have ever seen it burst like a rocket and lighted*
*up the whole heavens it was very grand indeed. There is*
*said to be a fine comet visible now but I have not seen it.*
*We propose looking for it tonight.*

*I have heard nothing more about coming home.*
*There is a possibility – no more.*

*Mrs. Crooke says to thank mother for her kind mes-*
*sages about her having to do housework. It is really a*
*shame and once in a while I try to give her a little help*
*particularly on Sundays.*

*Everything is moving along very swimmingly. I*
*wish you were here to dash off some letters for me.*

*I have pretty much made up my mind not to go*
*to S.A. though I do not at all know what I will do*
*after this year.*

Back at Lake City, Eben returned to work for the Crookes.
The newspaper quoted him in a brief comment on July 16, that

a "splendid streak" of gray copper had been opened on the sixth level of the Ule Mine. The smelter was also running, primarily on ore from the same mine. [19] Meanwhile, he wrestled with what he should do. The next letter finally developed what Olcott planned..

[July 22nd, 81]

Tonight Mr. Marshall called upon me. He was expected yesterday and I went down town and took supper there expecting to meet him but he did not arrive till today. I was just starting down to see him when he walked up. I was not over taken with him in the first place he came up here with two fellow travelers who are always an abomination. We do not like visitors as they got themselves into a little trouble. The man is close and belongs to a close [indecipherable] Co in Boston of which W. F. Durant founder of Wellsville ? [20] Female College is President & E.A. Hanford of Cambridge is Vice President while Mr. Marshal is Treasurer. He is prepared to make an engagement with me at $8000 and 2% of profits which they expect to be $500000 per annum but of course this is uncertain.

He wants me to go at once to Mexico and examine the property but I cannot do this before middle of Sept. and maybe not then. What I think would be most prudent would be not to commit myself till Sept. and try and hold him off till then and then go to see for myself whether I would engage for 1 yr. I don't mind the country and my Spanish would certainly come in play nicely.

The place is in Sonora 220 miles from Guaymas, on Gulf of California & 120 miles from nearest R.R. Now when A. S. and S. F. is all completed it will be only

*30 miles to nearest station and within 6 days of NY. I
would go in a minute to see the property if I could.*

*I have not time for more now. I will see Marshall
tomorrow and write on Sunday. He will be here till
Monday A.M. He is rigid observer of Sunday which will
be pleasant we will go to church together. He seems to
want me and I may make some sort of deal with him.
Mother need not worry. I haven't gone to Mex. yet and
it is just as likely I will not.*

*Excuse sealed enclosure. In the summer time there
is no telling who may open letters.*

It is interesting that he worried in the "summer time" about
who might open letters. Why he was concerned is not known. In
his next letter, Eben continued describing his "courtship" about
a new job.

*[July 24th, 1881]*
*Your letter of last Sunday reached me yesterday
and I am glad it did for today we have no mail owing
to a wash out just East of Alamosa.*

*Yesterday A.M. tore down to see Mr. Marshall in
P.M. I sent him my horse as I was busy he went to the
mines. I ordered a boy to wait for it and bring it up but
it did not arrive before I wanted to go down in the
evening. Wm. and I started on foot but met the boy just
below here I offered Wm the horse but as he refused he
said he would ride behind but instead I told him to get
in the saddle and I would mount behind him which I
did. Joe tried his level best to get me off and then would
not go for a few minutes but soon we got him started and
then he went on a good canter all the way to and*

*through town it was most ridiculous to see us and we cre-
ated some merriment in town.*

*Tomorrow A.M. I start for the North Star Mine to
do a little work there. I do not know how long I will be
away but continue to direct letters here. The decision
about the North Star has just this minute been made
and I have much to do. I shall go to town this evening
and go to church with Marshall. I have a strong idea
that I will go to Mexico. I intend to get out of this shortly
just as soon as I can leave in good graces. I telegraphed
Frank yesterday to know if he would go to Mexico for 3
months, two before I go and 1 month after I am there.*

*I cannot discuss reasons today I must stop at once
with much love to all.*

Eben did not wait that long to make a decision. With the
situation with the Crookes and the North Star no better, he was
actively looking to move on. His use of the telephone must have
been a new experience because he talked about it so much in his
next letter.

*N.S. Mine*
*[July 28th, 81]*

*As usual I am in a rush my eyes gave out last
night before my work was 1/2 done and now I cannot
stop to write. I presume Frank Rhodes left Sandy yesterday for Sonora Mexico to be there 2 weeks till I can come
that is if I conclude to. I have a strong inclination to go.
Don't you people worry about it. Telegrams which I
expect in 3 or 4 days from N.Y. will decide the matter. I
had a great time fixing up with Marshall. I had to
hurry off here before matters were settled. I used the telephone at 3 places before I reached here and 3 times here*

*to try and straighten things up. The telephone works well between Silverton and Lake 33 miles.* [21]

*We are having a fine snow storm* [22] *just now, heavy rain yesterday grand mt. effects all the time. I hope to be back in Lake inside of 2 weeks but may not. My correspondence will be much behind hand and if anything important comes up telegraph to Lake City and the message will be repeated by telephone to Silverton.*

*The trip over here was gorgeous my horse is much better than he was last year in fact a jewel.*

The last letter was sent from the North Star Mine on Pheme's birthday, as Eben excitedly looked forward to a new adventure, even trying out some of his Spanish. It was not all excitement, however, as his letter indicates. Olcott still worried about going, but some unstated reasons pushed him.

*[July 29th 81]*
*You could not have had a more exquisite day for your birthday even if you were not so fortunate as to be on top of the Rocky Mountains. It is the first day since I have been here that it has not rained although we had a delightful little hail storm this evening. After supper tonight I dropped down into the valley nearly 1000 ft. and had quite a walk back. The sky was a little overcast and it began to hail a little but the clouds lifted to allow of a glorious sun set which made mountains clouds and sky variously brilliant then sinking it left the Western horizon beautifully lines with black clouds behind me of which the moon shown bewitchingly. Mountains, mountains everywhere pointed, round, crested butte-like, slanting, irregular, as stiff as a wooden cone in fact*

*every conceivable shape that can be imagined has been thrown up in the wonderful platonic upheaval.*

*I am luxuriating in directing 20 men without a foreman so it keeps me pretty busy and in the evenings I miss John Cruse. I have not been to town since Tuesday and don't expect to go down till Sunday. I had hoped for letters by our man today but there were none of any importance. I should like to have one from Frank it is nearly time that one should be here since my telegram to him about going to Mexico. I don't want to go contrary to your advice or to worry you but I have very strenuous reasons for wishing to go. The country is no more dangerous than most of our western states and territories. There are hundreds of Americans there and as the R.R. extends thither the number is rapidly increasing. The mines have a splendid reputation and have been worked 50 yrs. and consequently well proved. The climate is excellent and the distance from home nothing to speak of. My experience with people and mills in S.A. as well as my speaking the lingo makes me valuable to them and I believe I can do more towards building up my reputation there than in the U.S. The parties I guess are good and have money on hand to push through their work.*

*I hate to change but it seems necessary for a M. E. to do so pretty often. In one sense this is very valuable and desirable for it enlarges our scope as consulting engineers when we rise to that dignity and settle down in N.Y. with a swell office and other blessings. (Castles in the air ?no es verdad?)*

*I don't know when I will get off from this connection but probably by Sept. 15th and then rush to Mexico decide on the mill etc. and back to San Francisco. Such at least seems likely to be the way things will turn out. I wish I only knew positively.*

*The Congregational Minister in Silverton is a pleasant little man and he is also County treasurer. Is not that a happy combination for such a place? He is highly respected in both places.* [23] *Marshall tried to hire Debble to go to Mexico as my mining captain. The whole Co. are very strict as regards all religious observances and [included] the stipulation in my agreement (verbal) that there shall be positively no work due on Sunday. The Co. owns 500 head of oxen and 40000, acres agricultural land so I am likely to be farmer as well as miner. The mines are near Banamechi 5 miles W of the Sonora river in Sonora 262 N.E. of Guaymas 120 miles S of the Arizona line.*

*If you were only here tonight to mend a sad rent in my trousers I would be comparatively happy. There might be some slight necessary pleasures in your presence too trivial to mention in comparison with thoughts of a [??]. When I came over here I carry scarcely anything except what I wear.*

*I had a visit yesterday from a man who made $60,000 last year who promised to do an errand for me in town. He wrote me today. "I send six which is all you can youse now I suppose I go to animas forks* [24] *and will bee in lake on Monday."* [25] *Don't you wish you were rich? … My men are quiet, thank fortune and I must very soon to bed myself. You might look in upon us; my little box of a room is comfortable enough but to see the room where 20 men sleep eat and live besides our storing away all our provisions you would think it's a steerage deck of a ship a neat abode. Hams and Beans are strung on one side of the room bags and coffee sugar tea decorate the beams over head shovels and picks occupy one corner of them, 5 double bunks, 3 tiers high take up considerable*

*room and the dining table and stove all the rest & boxes*
*etc are stuck around wherever there is a possible resting*
*place When men's boots get piled up around the stove to*
*dry and pretty often begins to smoke a vile pipe. I will*
*stop you can imagine the rest.* [26]

*Maybe a large house to oneself and men living in*
*little adobe dwelling at respectful distances may be an*
*improvement. Still what miner minds such little incon-*
*veniences and if you think I don't enjoy my meals in that*
*self same room you are very much mistaken. The con-*
*trast between the table and that of Mrs. Crooke is strik-*
*ing I can assure you.*

Silverton's *La Plata Miner* covered North Star develop-
ments that summer of 1881 far better than the *Silver World*. The
mine, after all, sat in its mining hinterland, even thought the
Crookes and the company headquartered in Lake City.

In March, the paper reported that the company expects "to
push work more vigorously than ever before." A month later, on
April 16, it was understood that the foreman would arrive in a
"few days" to start up work. About "10,000 lbs. of supplies are at
the mine," having been taken up last fall as part of Eben's efforts.
It did not quite work out as planned, and not until July 30 was the
*Miner* "pleased to announce" work had commenced.
Superintendent Olcott had a force of twenty men working at the
mine. "Here-to-fore" the North Star had been "one of the largest
high grade ore" producing mines in the country. Cheered the editor, "we look for flattering results from working on the property."

That would not be this season. The paper did not give up
hope, though. The August 20 issue carried a short piece about
shipping a pack train load of ore "every day," with the "mine
looking better than ever." The ore according to the report ran
"200-400 ounces" to the ton. One is reminded of Mark Twain's
quip about his Nevada mining days, "We lived in a little cabin
and cooked for ourselves; and altogether it was a hard life,

though a hopeful one—for we never ceased to expect fortune and a customer [purchaser] to burst upon us someday."

The Crookes might have held on to that hope, but Olcott had had enough. The *Silver World* traced Even's last days. On August 6, it announced L. P. Burrows had been appointed general manager of the North Star. Then on Tuesday August 9, Olcott left Lake City for New York. The *Miner,* September 17, marked the end of his San Juan career: "Prominent mining engineer," E. E. Olcott, left New York City to "take charge of the St. Lenena Mine at Sonora, Mexico."

After leaving the North Star, Olcott wrote a letter from Mexico to James Elwood. He received this most fascinating reply (September 25, 1881). Elwood still worked at Lake City, and matters were not going well with either the North Star or the company. A "band of experts" had just "come and gone leaving a pleasant memory." He had not heard anything about their reports, however Olcott's "actions here furnished an interesting topic." What those were both men knew, so did not elaborate. The North Star was rumored to be up sale, but that was denied in Lake City. The Crookes were pointing fingers over unexplained matters, and an unidentified officer was his "usual or customary S. of a B." self.

Elwood thought it "just as well that you [Eben] left when you did," and then added, without elaborating, "as it was far more manly and more in keeping with professional dignity." He was glad Eben was in good spirits and "not sighing for the good (!) old days of the San Juan in general the Crookes in particular!"

In that letter lie the reasons for Olcott's leaving, combined with some of the issues and personalities he had already mentioned. The hints prove most tantalizing. It almost seemed like the enterprise was teetering on the brink, which the future would determine it was indeed doing. Anyway, Olcott's days in the San Juans were over. He had learned a great deal about mining, particularly mining in an undeveloped, underfinanced, isolated region with a mine at a high elevation and owners who risked much, expected more, and found so few financial returns.

The church-going Eben Olcott would probably have agreed with Martin Luther, whose father owned mines. Luther wrote these words in rejecting shares in a German mine, because, he affirmed, "I would have no luck in mining."

> *Satan deludes many in mines, making them think*
> *they see great store of copper and silver where there is none.*
> *If he can bewitch men in full daylight above ground, he*
> *can do so much more in a subterranean mine.*

These sentiments, true more times than not, remind one of the Crookes, not the practical, experienced mining engineer now on his way to Mexico.

1. The small hamlet that grew up around the smelter one mile from Lake City was where Olcott spent that winter..

2. *Silver World*, January 29 & February 5, 1881.

3. Euphemia was the name of Olcott's mother. It is probably also the full name of Pheme.

4. The *Silver World* reported on February 12, 1881, that 2,000 letters had been collected at the scene of a mail robbery, many showing the effect of exposure to snow and water. The editor thought the robbers "did not realize more than $100. Poor pay."

5. Rev. A. D. Drummond. Services were at ll.

6. The Ute/Ule mine was two miles from Crookesville, three miles from Lake City on Henson Creek. Ulay or Ule—the spelling varied depending on the writer.

7. A small town among the hills near Utica.

8. John Debble was the acting pastor of the Presbyterian church, which had only 7 pm Sunday services. *Silver World*, March 26, 1881. Olcott spelled it Deeble, the newspaper Debble.

9. Episcopal church at Lake City.

10. *Silver World*, April 2, 23, & 30, 1881.

11. Lake City was having troubles with robberies. The *Silver World*, March 26, 1881 had noted that "highwaymen for the fourth time within six months waylaid and robbed the mail and express coach."

12. Former union general James Garfield had become president in March. New York politician and United States senator, Roscoe Conkling, dramatically resigned from the senate in a dispute with Garfield over patronage. Olcott got his wish, Conkling was not reelected.

13. *Silver World*, May 28, 1881.

14. Olcott might have heard much about him, few others have!

15. Echo City, on the Union Pacific Railroad near Ogden, in 1868 had businesses, "nymphs du grade," and fifty flimsy structures as the construction crews arrived. The excitement proved brief, leaving the place with the name Echo.

16. This might be Charles L. Strong, who earlier had been one of the "remarkable" superintendents on the Comstock. Strong had been involved with George Hearst and others in the famous Gould & Curry mine. In 1878, he managed the White & Shiloh Mine in Lander County. Grant H. Smith, *The History of the Comstock Lode* (Reno: Nevada Bureau of Mines, 1980 reprint), 39 & 84-5. Olcott was, at the moment, in Galena, Lander County, Nevada.

17. Mining in the Battle Mountain district started in 1864; the peak of its mining activity came eight years later. Low-grade copper and silver were the chief products, with a little lead and gold.

18. Olcott is teasing. The Utes did not poison arrows.

19. *Silver World*, July 16, 1881.

20. Olcott had a question mark here. No Wellsville Female College turned up in a search although a town did.

21. The *Silver World* had discussed the telephone in the late winter of 1881. The line to Silverton was erected in the spring and early summer. Alexander Graham Bell invented the telephone in 1876, and the mining West quickly adopted the invention as a way to overcome isolation and distance.

22. A July snow storm was not particularly unusual. It can snow any month in the high elevations of the San Juans.

23. Harlan Page Roberts was the minister. After nearly a year of volunteer construction work, the church was completed in July. The First Congregational Church was dedicated on the tenth, so Olcott attended one of its earliest services. The building is still standing. See Nossaman, *Many More Mountains*, vol. 111, 30-34.

24. A small camp located at the head of the Animas River. At 11,200 feet, it was one of the San Juans' highest and despite a peak summer time population of 200-300 (a quarter of that in the winter), managed to have its own newspaper, the *Animas Forks Pioneer.*

25. The cultured and educated Olcott does not usually let his feelings about some of his companions reach paper. There are hints throughout his letters, however, that he found some of them less than appealing.

26. Such crowded conditions were not unusual in early mining districts, particularly at an elevated spot such as the North Star.

EPILOGUE

*A Career Well Spent*

# EPILOGUE

*A*fter weighing all the factors both pro and con, along with some soul searching, Eben Olcott decided to go to Mexico. He spent the next four years as superintendent of the St. Helena Gold Mining Company in Sonora, Mexico. Four years "of strenuous work under untoward conditions" in developing the property and building a plant proved enough for the venture. Now married, he had a further reason to come back to the states.

Olcott then returned to his consulting practice, with an office in New York City that he had dreamed about years before in the San Juans. His work evolved into endeavors in both North and South America. With the advantage of being fluent in Spanish, Eben traveled to Peru, British Guiana, Colombia, New Brunswick, Ontario, Mexico, and throughout the United States during the next decade. A 1890-91 examination, for instance for the Peruvian Exploration Syndicate of London took him up to more than 17,000 feet in the Andes to report on gold and copper deposits. Meanwhile, in 1884, he married Kate Van Santvoord and settled down in New York City. They became the parents of three sons and a daughter.

His full-time mining career came to a close in 1895 with the death of his brother-in-law who had been general manager of the Hudson River Day Line, a system of passenger steamers operating between new York City and Albany. Started by his wife's father, the company was running only two steamers when Olcott became general manager. He increased that to seven larger steamers after he assumed the presidency of the firm in 1901. One of his obituaries noted: "No detail that might contribute to the success of the business was too small to receive his personal attention. He made himself familiar with every feature of the boats, from boiler and engine rooms to the Commissariat Department."

Eben Olcott never completely forsook mining. He reorganized his consulting engineering firm in 1896 with partners and continued for years to "give part of his time, in an advisory capacity" to its activities. In 1901, while president of the American Institute of Mining Engineers, he led a visit to Mexico and convened a meeting at the Mexican National School of Engineering. His command of Spanish came in handy, as he gave a "very flowery address" at the opening session. After praising Mexican mining contributions, Olcott asserted that the United States should have founded its mining laws on Mexico's. "If we had done so, the vexatious perplexities of the 'apex-question' would have been avoided, and the rational, understandable, rectangular claims would have been adopted. [1]

Praised for his "forceful, impressive personality," his "integrity, large humanity, kindly courtesy, and never-failing helpfulness," he lived out his life in New York City dying there, on June 5, 1929. The American Society of Civil Engineers hailed him as having "the affection, loyalty, and respect of all who were brought into contact with him" in his two careers as a mining engineer and a New York businessman.

In his papers, there exist a few references to the North Star Mine after he left, but as far as is known, Olcott never returned to the San Juans. The mine went on to a productive, if not spectacular, career. The 1883 report of the *Director of the Mint*, for example, noted, "the past summer there have been taken out and shipped 550 tons of ore, that run 100 ounces silver, one-half ounce gold and 40 per cent lead." A summary of the North Star Mine production from July 1, 1883 through November 1, 1897 gave the gross value as $1,294,892.55, almost entirely from silver.

A description of the mine, written about this time, stated that the North Star Company had worked the mine for "some 20 years," with ore running from "60 to 300 ounces." The main shaft was down about 800 feet with twelve miles of workings and five levels within the mine carrying ore and drainage. The "number of levels running to the outside will not greatly cheapen the handling and exploiting for mineral." Most telling however were these comments:

> *The property is in complete order and only awaits capital.*
> *What this great property wants is further exploiting, together with a tram to Cunningham Gulch. This would greatly economize the severe charges of transportation. And make possible a liberation of tens of thousands of dollars now lying useless on the dumps. A mill located on Cunningham gulch would be a great desideratum.*

Interestingly, a mill located three and a half miles from the mine, was the property of John J. and Robert Crooke.[2]

Isolation and altitude continued to cause problems, even with a lower portal, and, by the turn-of-the-century, the North Star had receded from the list of important producing mines. Hope springs eternal though, and, in the twenty-first century, there still exists interest in the old property. Why? A late 1890s report predicted.

> *This property is located in the immediate vicinity of several profitable gold mines lying some twelve hundred feet lower. The presumption is that with a depth to reach the level of these properties, gold will come in with the North Star ore.* [3]

That depth would never be attained and the theory tested.

San Juan County, however, went through its greatest mining boom from the 1890s, and on into World War I, averaging production of nearly two million dollars per year in gold, silver, lead, and copper. One of the leading mines for much of this period was the other North Star over on Sultan Mountain. While some of the small mining camps Olcott knew declined and disappeared, the county seat, Silverton, prospered as never before, topping 2,000 population in 1910.

Unfortunately for Hinsdale County, none of the same could be said. As the report of the Director of the Mint (1884) observed, "Hinsdale County's production has fallen off considerably. ...with the closing of the Crooke Mining & Smelting Co." Never a major producing district, it reported only $287,000 for 1885. After a short-lived rally in 1890-91, Hinsdale's production slipped under even that. Lake City's population followed the same trend, dropping to 405 in 1910.

As indicated, the Crooke brothers' operations did not survive. Despite being given a "great deal of credit" in the *Engineering and Mining Journal* ( January 17, 1880) for their pioneering smelting efforts, the article went on to explain they had gone through years of costly experiments to find a treatment process. Still, not all the problems were resolved, and this led to a temporary closure. The *Silver World* told its Lake City readers, April 7, 1883, that the cost of fuel, flux and labor, plus the expense of hauling ore from the mines, had caused a temporary closure. The editor also blamed unspecified company policy for this situation.

Nor did some of their mines pan out as anticipated particularly the Ute and Ulay. The result hampered operations until the closing of activity in November 1883. The *Lake City Mining Register* (November 2, 1883) praised "Uncle" John Crooke as the "chief engineer of the city's prosperity." It blasted the "cheerful idiot who did not see a faint idea of what one man's conceptions, executions, ambitions and grit are worth." Other sources blamed "troubles among themselves" for the company's trials.

High expenses, low grade ore, debt, perhaps some poor leadership or choices, and a local district that did not produce as expected doomed their efforts. The company's assets were sold at a public auction in May 1886 in New York City. Once again hope blossomed. The *Engineering and Mining Journal* reported, on June 19, 1886, that as "soon as proper arrangements can be perfected," the new owners could then work again. Primarily, this meant the Ute and Ulay Mine.

The San Juans, where Olcott worked, eventually emerged as one of the United States's, and even the world's, great mining districts. Ouray and Telluride, joined by an invigorated Silverton at the turn-of-the-century, pushed the San Juans into the mining headlines with outstanding production. By 1923, the glory days had receded, but the district had produced more than twenty-two million ounces of gold and twenty-eight million ounces of silver. Adding in the value of gold and silver, the production of zinc, copper and lead created a total amount of $70,381,000. This mineral treasure box also produced coal, tungsten, uranium, and, along its southern borders, oil and natural gas.

In his classic *Many More Mountains* (volume 3), Silverton historian Allen Nossaman pays a fitting tribute to Eben Olcott.

> *He was perhaps typical of a number of refined and well-educated men who engaged in a confrontation with the lofty San Juans—learning and contributing much, but finally deciding to leave— and his surviving letters contain so many valuable observations on both mining and community life during the 1880-81 winter that they deserve more than passing attention.*

Olcott was one of the pioneering mining engineers in the San Juans, who helped make possible its early development and laid the foundation for those who came after. His stay had been brief, but it became part of a larger whole—settlement and growth of the San Juans, of Colorado, and of the Rocky Mountain West.

SAN JUAN GOLD: *A Mining Engineer's Adventures, 1879-1881*

1. Nothing was more contentious in American mining law than the apex question. Defined in the 1872 mining law as the highest point of a vein, it was a lawyer's delight, because the mine that held the apex controlled the vein across the side-lines. Mining engineers and geologists of the late nineteenth century could not easily define that point, hence expensive lawsuits followed. Noted mining engineer James Hague contended that the apex law brought "forth nothing but confusion and contention, with everlasting and costly litigation."

2. Both reports are in the Olcott Papers, New-York Historical Society.

3. This is probably part of a report of mining engineer Robert Peele who examined the property in July 1899. If so, he was referring to the Iowa, Dives, and Silver Lakes Mines. He discussed, in his field notes, transportation problems, and raised the question of "how long before tramway would pay for itself?"

# FURTHER READING

*T*he San Juans, as one of North America's major mining districts, have attracted their share of scholarship. Allen Nossaman's three volumes, *Many More Mountains* are a bonanza of information on every conceivable topic. P. David Smith's *Mountains of Silver* examines the Red Mountain district. John Marshall with Zeke Zanoni, *Mining the Hard Rock in the Silverton San Juans* covers yesterday and today. Duane A. Smith's *Song of the Hammer and Drill: The Colorado San Juans 1860-1914* gives the overview for those decades. T. A. Rickard, *Across the San Juan Mountains* takes the reader back to the turn-of-the-century. *Red Mountain*, David Lavender's fiction account of opening the San Juan is historically accurate, insightful, and a crackling good tale. Silverton and Lake City have attracted their share of short histories and pamphlets on a variety of topics. All of these will provide further things to think about and more books and places to go and to read/research.

To really gain a feel for the Olcott's era look into the Lake City and Silverton newspapers. They are invaluable sources. The *Engineering and Mining Journal* carries San Juan articles throughout the last three decades of the nineteenth century and to a lesser degree so does Denver's *Rocky Mountain News* and Durango's variety of newspapers. Charles Henderson's *Mining in Colorado* provides production records and county by county overviews.

The two collections of Olcott materials at the New-York Historical Society and the American Heritage Center at the University of Wyoming tell much about the man and his work beyond the scope of this study. Here lies the real heart of historic research.

Nothing can be more fun, exciting, and rewarding than to pursue your own journey into the past. The San Juans offer such a colorful tapesty that there is no better place to start.

African Americans, 101-02
Alamosa, 71
Altitude: impact, 33; 45 (n9 & 10)
American House, 30, 44 (n3)
Animas Forks, 116, 120 (n24)
Antelope Springs, Colorado, 32, 45 (n8)
Apex, 25 (n23), 124, 128 (n1). *See also United States Mining Law*
Argo Smelter, 11, 24 (n7)
Aspen, Colorado, 10


Baker, Charles, 1
Baker's Park, 1
Banamechi, Mexico, 116
Bassick Mine, 15
Battle Mountain, Nevada, 109, 120 (n17)
Beaver County, Utah, 18, 25 (n20)
Boulder, Colorado, 11, 24 (n5)
Boulder Gulch, 5
Burrows, L. P., 118


California, 108
Canon City, Colorado, 12, 21, 26 (n26)
Caribou, Colorado, 11, 24 (n5)
Central Presbyterian Church, Denver, 11, 24 (n4)
Charpiot's Hotel, 11, 14, 20, 24 (n6)
Chrysolite Mine, 81, 92 (n 13 & 15)
Clarendon Hotel, 17, 25 (n16)
Cleora, Colorado, 21, 26 (n27)
Clinton, New York, 102, 119 (n7)
Colorado, 3, 10, 21
Communication: problems, 15
Conkling, Roscoe, 120 (n12)
Connor, William, 44 (n3)
Crime: mail robberies, 100, 104-05, 119 (n4&11); murder, 75-76; vigliante
    committee, 79-80, 92 (n11)
Crooke, Charles, 5
Crooke, Henry, 105

SAN JUAN GOLD: *A Mining Engineer's Adventures, 1879-1881*

Crooke, John, 4, 29, 77, 104, 107, 125; North Star Mine, 73, 74-75; praised, 126
Crooke, Miss, 102-03
Crooke, Mrs. John, 99-100, 101, 110
Crooke, Lewis, 5, 76, 77
Crooke, Robert, 5, 125
Crooke Mining and Smelting Company, 13, 14, 77, 98
Crooke Smelting & Concentrating Works, 54 (drawing)
Crookes (Family), 36-37, 39; criticism of, 102; evasive, 82; fail, 126; significance, 77; troubles, 118
Crookesville, Colorado, 76, 91 (n2), 97, 119 (n1 & 6)
Cruse, John, 115

Darley, George, 45 (n12)
Debble, John, 103, 106, 107, 119 (n8)
Dekay, Mrs., 81, 92 (n15)
Del Norte, Colorado, 2, 31, 32, 43 (n6)
Denver, Colorado, 10-12
Denver & Rio Grande Railroad, 22, 26 (n30), 81
Denver, South Park & Pacific Railroad, *See South Park Railroad*
Dodge, David, 14, 24 (n10)
Dogs, 89
Drummond, P. D., 119 (n5)
Durango, Colorado, 26 (n30), 81, 82

Echo City, Utah, 108, 120 (n15)
Elwood, James, 118
Engineering and Mining Journal, 31, 44 (n4)
Episcopal Church, Lake City, 100-01
Eureka, Nevada, 20, 25 (n23)

Farish, John, 17, 18, 26 (n19)
Flowers, 40-41, 45 (n18)
Flux, 78, 92 (n6)
Foote, Arthur, 24 (n2)
Foote, Mary Hallock, 10, 24 (n2)
Frisco, Utah, 18-19, 25 (n20)

Skems, Sam, 79
Skiis, 56 (drawing), 84, 92 (n18)
San Juan & New York Smelter, 92 (n8)
Snow, 5, 56 (drawing), 73, 80-81, 84, 85, 86, 87, 91; July storm, 114, 120 (n22)
San Juan & New York Smelter, 92 (n8)
South Park Railroad, 20
Stagecoach, 19, 21-22; accident, 30-31, 32-33
Stevens, William, 25 (n25)
Stony Pass, 2
Stratton, Winfield Scott, 4
Strong, Charles, 108-09, 110, 120 (n16)
Summitville, Colorado, 44, 46 (n23)

Teitsworth, W. P., 17, 24 (n14)
Telephone, 113-14, 120 (n21)
Travel, 12, 71; railroad, 20-21, 22-23; stage, 21-22, 30-31, 32-33
Twain, Mark, 117-18

Ule (Ulay) Mine, 77, 104, 111, 119 (n6), 126
Uncompahgre Peak, 34 (n11)
United States Mining Law, 78, 92 (n9). *See also Apex*
Urbanization, impact 2. *See also Denver; Lake City; mining camps; Silverton*
Ute Mine, 77, 126, 127
Utes, 1, 4, 22, 26 (n29), 109, 120 (n18)

Van Santvoord, Kate (wife Eben Olcott), 123

Wagon Wheel Gap, Colorado, 31, 44 (n7)
Walker, William , 92 (n16)

Walker House, 53 (photo), 82, 83, 92 (n16)
Weather, 87, 97, 110; beautiful, 15-16, 71, 73, 88, 105; cold, 89, 90, 98; rain, 114; wind, 85, 86. *See also nature; scenery*
Westcliffe, Colorado, 24, (n12)
Winter, 102-03. *See also snow*
Wisconsin Central Mines, 18
Women, 21, 61; assayers, 43-44, 46 (n22 & 24); traveling, 74